Essential Physics Part 1

Essential Physics Part 1

RELATIVITY, DYNAMICS, GRAVITATION & WAVE MOTION

Frank W. K. Firk
Professor Emeritus of Physics
Yale University

ISBN – 13: 978-1499396041
ISBN – 10:149939604X

CONTENTS

PREFACE

Throughout the decade of the 1990's, I taught a one-year course of a specialized nature to students who entered Yale College with excellent preparation in Mathematics and the Physical Sciences, and who expressed an interest in Physics or a closely related field. The level of the course was that typified by the Feynman *Lectures on Physics*. My one-year course was necessarily more restricted in content than the two-year Feynman Lectures. The depth of treatment of each topic was limited by the fact that the course consisted of a total of fifty-two lectures, each lasting one-and-a-quarter hours. The key role played by *invariants* in the Physical Universe was constantly emphasized. The material that I covered each Fall is presented, almost verbatim, in this book.

The first chapter contains key mathematical ideas, including some invariants of geometry and algebra, generalized coordinates, and the algebra and geometry of vectors. The importance of linear operators and their matrix representations is stressed in the early lectures. These mathematical concepts are required in the presentation of a unified treatment of both Classical and Special Relativity. Students are encouraged to develop a "relativistic outlook" at an early stage. The fundamental Lorentz transformation is developed using arguments based on symmetrizing the classical Galilean transformation. Key 4-vectors, such as the 4-velocity and 4-momentum, and their invariant norms, are shown to evolve in a natural way from their classical forms. A basic change in the subject matter occurs at this point in the book. It is necessary to introduce the Newtonian concepts of mass, momentum, and energy, and to discuss the conservation laws of linear and angular momentum, and mechanical energy, and their associated invariants. The discovery of these laws, and their applications to everyday problems, represents the high point in the scientific endeavor of the 17th and 18th centuries. An introduction to the general dynamical methods of Lagrange and Hamilton is delayed until Chapter **9**, where they are included in a discussion of the Calculus of Variations. The key subject of Einsteinian dynamics is treated at a level not usually met in at the introductory level. The 4-momentum invariant and its uses in relativistic collisions, both elastic and inelastic, is discussed in detail in Chapter **6**. Further developments in the use of relativistic invariants are given in the discussion of the Mandelstam variables, and their application to the study of high-energy collisions. Following an overview of Newtonian Gravitation, the general problem of central orbits is discussed using the powerful method of [p, r] coordinates. Einstein's General Theory of Relativity is introduced using the Principle of Equivalence and the notion of "extended inertial frames" that include those frames in free fall in a gravitational field of small size in which there is no measurable field gradient. A heuristic argument is given to deduce the Schwarzschild line element in the "weak field approximation"; it is used as a basis for a discussion of the refractive index of space-time in the presence of matter. Einstein's famous predicted value for the bending of a beam of light grazing the surface of the Sun is calculated. The Calculus of Variations is an important topic in Physics and Mathematics; it is introduced in Chapter **9**, where it is shown to lead to the ideas of the Lagrange and Hamilton functions. These functions are used to illustrate in a general way the conservation laws of momentum and angular momentum, and the relation of these laws to the homogeneity and isotropy of space. The subject of *chaos* is introduced by considering the motion of a damped, driven pendulum. A method for solving the non-linear equation of motion of the pendulum is outlined. Wave motion is treated from the point-of-view of invariance principles. The form of the general wave equation is derived, and the Lorentz invariance of the phase of a wave is discussed in Chapter **12**. The final chapter deals with the problem of orthogonal functions in general, and Fourier series, in particular. At this stage in their training, students are often under-prepared in the subject of Differential Equations. Some useful methods of solving ordinary differential equations are therefore given in an appendix.

The students taking my course were generally required to take a parallel one-year course in the Mathematics Department that covered Vector and Matrix Algebra and Analysis at a level suitable for potential majors in Mathematics.

Here, I have presented my version of a first-semester course in Physics — a version that deals with the essentials in a no-frills way. Over the years, I demonstrated that the contents of this compact book could be successfully taught in one semester. Textbooks are concerned with taking many known facts and presenting them in clear and concise ways; my understanding of the facts is largely based on the writings of a relatively small number of celebrated authors whose work I am pleased to acknowledge in the bibliography.

Guilford, Connecticut

February, 2000

I am grateful to several readers for pointing out errors and unclear statements in my first version of this book. The comments of Dr Andre Mirabelli were particularly useful, and were taken to heart.

March, 2003

1

MATHEMATICAL PRELIMINARIES

1.1 Invariants

It is a remarkable fact that very few fundamental laws are required to describe the enormous range of physical phenomena that take place throughout the universe. The study of these fundamental laws is at the heart of Physics. The laws are found to have a mathematical structure; the interplay between Physics and Mathematics is therefore emphasized throughout this book. For example, Galileo found by observation, and Newton developed within a mathematical framework, the Principle of Relativity:

the laws governing the motions of objects have the same mathematical form in all inertial frames of reference.

Inertial frames move at constant speed in straight lines with respect to each other – they are mutually non-accelerating. We say that Newton's laws of motion are *invariant* under the Galilean transformation (see later discussion). The discovery of key *invariants* of Nature has been essential for the development of the subject.

Einstein extended the Newtonian Principle of Relativity to include the motions of beams of light and of objects that move at speeds close to the speed of light. This extended principle forms the basis of Special Relativity. Later, Einstein generalized the principle to include accelerating frames of reference. The general principle is known as the Principle of Covariance; it forms the basis of the General Theory of Relativity (a theory of Gravitation).

A review of the elementary properties of geometrical invariants, generalized coordinates, linear vector spaces, and matrix operators, is given at a level suitable for a sound treatment of Classical and Special Relativity. Other mathematical methods, including contra- and covariant 4-vectors, variational principles, orthogonal functions, and ordinary differential equations are introduced, as required.

1.2 Some geometrical invariants

In his book *The Ascent of Man*, Bronowski discusses the lasting importance of the discoveries of the Greek geometers. He gives a proof of the most famous theorem of Euclidean Geometry, namely Pythagoras' theorem, based on the *invariance* of length and angle (and therefore of area) under translations and rotations in space. Let a right-angled triangle with sides a, b, and c, be translated and rotated into the following four positions to form a square of side c:

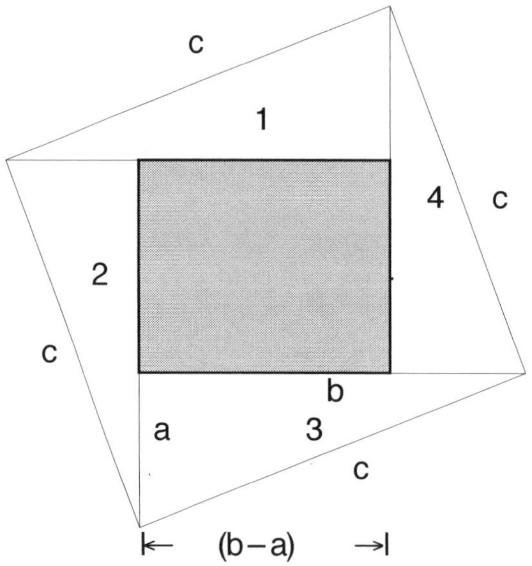

The total area of the square = c^2 = area of four triangles + area of central square.

If the right-angled triangle is translated and rotated to form the rectangle:

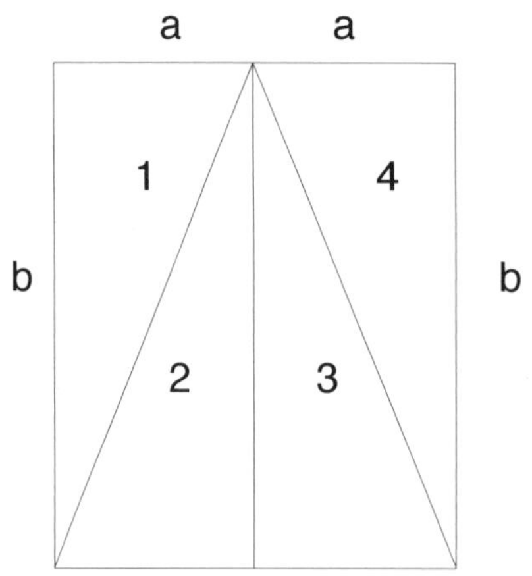

then the area of four triangles = 2ab.

The area of the shaded square area is $(b-a)^2 = b^2 - 2ab + a^2$

We have postulated the invariance of length and angle under translations and rotations and therefore

$$\begin{aligned} c^2 &= 2ab + (b-a)^2 \\ &= a^2 + b^2. \end{aligned} \qquad (1.1)$$

We shall see that this key result characterizes the locally flat space in which we live. *It is the only form that is consistent with the invariance of lengths and angles under translations and rotations* .

The *scalar product* is an important invariant in Mathematics and Physics. Its invariance properties can best be seen by developing Pythagoras' theorem in a three-dimensional coordinate form. Consider the square of the distance between the points P $[x_1, y_1, z_1]$ and Q $[x_2, y_2, z_2]$ in Cartesian coordinates:

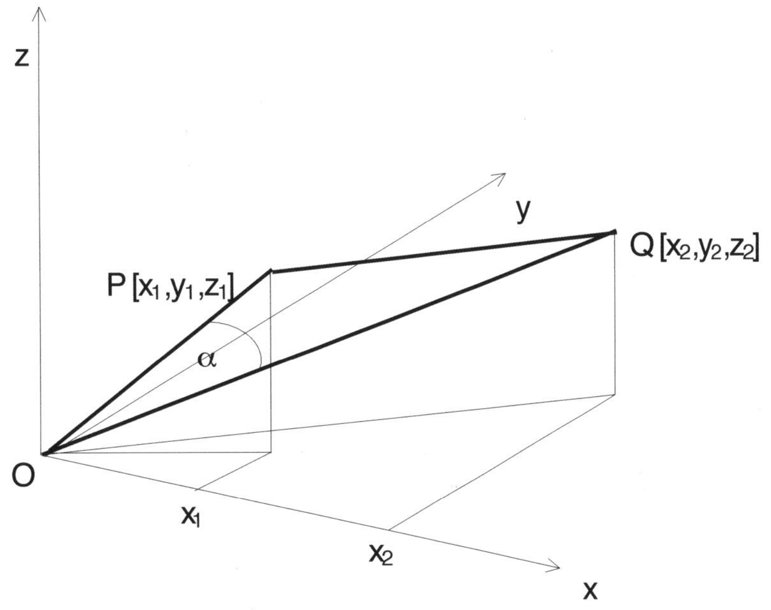

We have

$$(PQ)^2 = (x_2 - x_1)^2 + (y_2 - y_1)^2 + (z_2 - z_1)^2$$

$$= x_2^2 - 2x_1x_2 + x_1^2 + y_2^2 - 2y_1y_2 + y_1^2 + z_2^2 - 2z_1z_2 + z_1^2$$

$$= (x_1^2 + y_1^2 + z_1^2) + (x_2^2 + y_2^2 + z_2^2) - 2(x_1x_2 + y_1y_2 + z_1z_2)$$

$$= \quad (OP)^2 \quad + \quad (OQ)^2 \quad - 2(x_1x_2 + y_1y_2 + z_1z_2) \tag{1.2}$$

The lengths PQ, OP, OQ, and their squares, are invariants under rotations and therefore the entire right-hand side of this equation is an invariant. The admixture of the coordinates $(x_1x_2 + y_1y_2 + z_1z_2)$ is therefore an invariant under rotations. This term has a geometric interpretation: in the triangle OPQ, we have the generalized Pythagorean theorem

$$(PQ)^2 = (OP)^2 + (OQ)^2 - 2OP.OQ\cos\alpha,$$

therefore

$$OP.OQ\cos\alpha = x_1x_2 + y_1y_2 + z_1z_2 \equiv \text{the } \textit{scalar product.} \tag{1.3}$$

Invariants in space-time with scalar-product-like forms, such as the interval between events (see **3.3**), are of fundamental importance in the Theory of Relativity. Although rotations in space are part of our everyday experience, the idea of rotations in space-time is counter-intuitive. In Chapter **3**, this idea is discussed in terms of the *relative motion* of inertial observers.

1.3 Elements of differential geometry

Nature does not prescibe a particular coordinate system or mesh. We are free to select the system that is most appropriate for the problem at hand. In the familiar Cartesian system in which the mesh lines are orthogonal, equidistant, straight lines in the plane, the key advantage stems from our ability to calculate distances given the coordinates – we can apply Pythagoras' theorem, directly. Consider an arbitrary mesh:

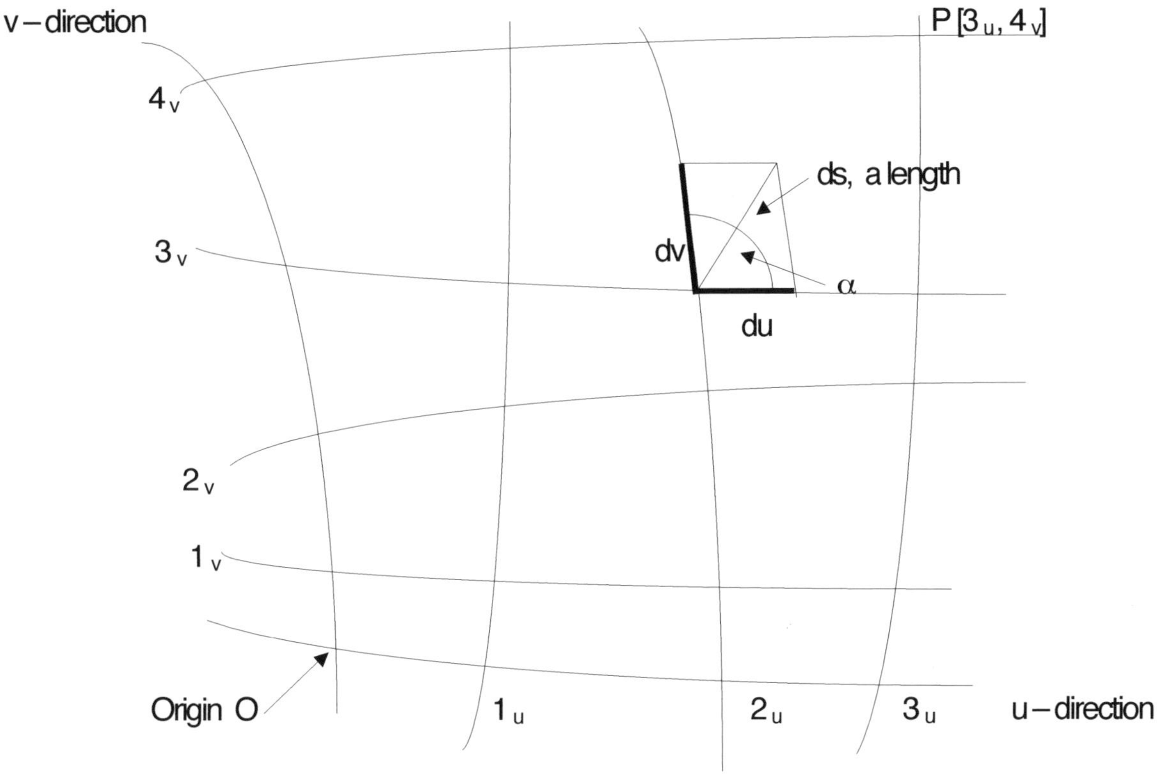

Given the point P [3 u , 4 v], we cannot use Pythagoras' theorem to calculate the distance OP.

In the infinitesimal parallelogram shown, we might think it appropriate to write

$$ds^2 = du^2 + dv^2 + 2dudv\cos\alpha . \quad (ds^2 = (ds)^2 \text{, a squared "length"})$$

This we cannot do! The differentials du and dv are *not* lengths – they are simply differences between two numbers that label the mesh. We must therefore multiply each differential by a quantity that converts each one into a length. Introducing dimensioned coefficients, we have

$$ds^2 = g_{11}du^2 + 2g_{12}dudv + g_{22}dv^2 \tag{1.4}$$

where $\sqrt{g_{11}}\,du$ and $\sqrt{g_{22}}\,dv$ are now *lengths*.

The problem is therefore one of finding general expressions for the coefficients;

it was solved by Gauss, the pre-eminent mathematician of his age. We shall restrict our discussion to the case of two variables. Before treating this problem, it will be useful to recall the idea of a *total differential* associated with a function of more than one variable.

Let $u = f(x, y)$ be a function of two variables, x and y. As x and y vary, the corresponding values of u describe a surface. For example, if $u = x^2 + y^2$, the surface is a paraboloid of revolution. The partial derivatives of u are defined by

$$\partial f(x, y)/\partial x = \text{limit as } h \to 0 \{(f(x + h, y) - f(x, y))/h\} \text{ (treat y as a constant)}, \tag{1.5}$$

and

$$\partial f(x, y)/\partial y = \text{limit as } k \to 0 \{(f(x, y + k) - f(x, y))/k\} \text{ (treat x as a constant)}. \tag{1.6}$$

For example, if $u = f(x, y) = 3x^2 + 2y^3$ then

$$\partial f/\partial x = 6x,\ \partial^2 f/\partial x^2 = 6,\ \partial^3 f/\partial x^3 = 0\,,$$

and

$$\partial f/\partial y = 6y^2,\ \partial^2 f/\partial y^2 = 12y,\ \partial^3 f/\partial y^3 = 12, \text{ and } \partial^4 f/\partial y^4 = 0.$$

If $u = f(x, y)$ then the total differential of the function is

$$du = (\partial f/\partial x)dx + (\partial f/\partial y)dy,$$

corresponding to the changes: $x \to x + dx$ and $y \to y + dy$.

(Note that du is a function of x, y, dx, and dy of the independent variables x and y).

1.4 Gaussian coordinates and the invariant line element

Consider the infinitesimal separation between two points P and Q that are described in either Cartesian or Gaussian coordinates:

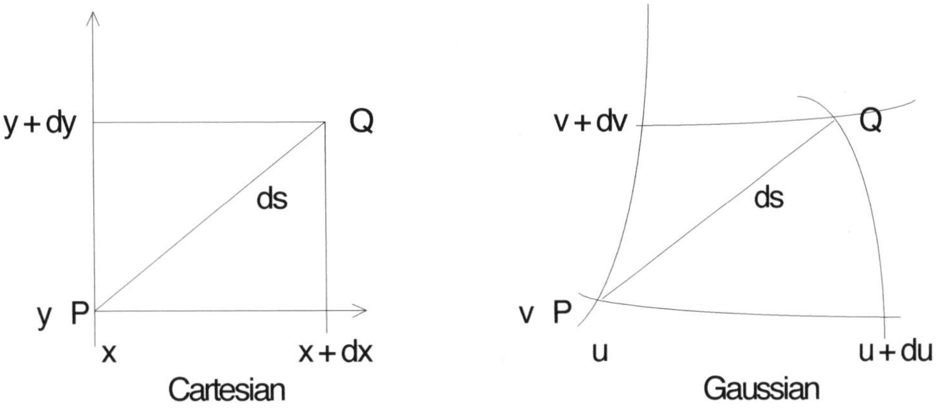

In the Gaussian system, du and dv do not represent distances.

Let

$$x = f(u, v) \text{ and } y = F(u, v) \tag{1.7 a, b}$$

then, in the infinitesimal limit

$$dx = (\partial x/\partial u)du + (\partial x/\partial v)dv \text{ and } dy = (\partial y/\partial u)du + (\partial y/\partial v)dv.$$

In the Cartesian system, there is a direct correspondence between the mesh-numbers and distances :

$$ds^2 = dx^2 + dy^2 . \tag{1.8}$$

But

$$dx^2 = (\partial x/\partial u)^2 du^2 + 2(\partial x/\partial u)(\partial x/\partial v)dudv + (\partial x/\partial v)^2 dv^2 \text{ and}$$

$$dy^2 = (\partial y/\partial u)^2 du^2 + 2(\partial y/\partial u)(\partial y/\partial v)dudv + (\partial y/\partial v)^2 dv^2.$$

We therefore obtain

$$ds^2 = \{(\partial x/\partial u)^2 + (\partial y/\partial u)^2\}du^2 + 2\{(\partial x/\partial u)(\partial x/\partial v) + (\partial y/\partial u)(\partial y/\partial v)\}dudv$$

$$+ \{(\partial x/\partial v)^2 + (\partial y/\partial v)^2\}dv^2$$

$$= g_{11}\, du^2 + 2g_{12}dudv + g_{22}dv^2 . \tag{1.9}$$

If we put $u = u_1$ and $v = u_2$, then

$$ds^2 = \sum_i \sum_j g_{ij} du_i du_j \text{ where } i, j = 1,2, \tag{1.10}$$

(a general form in n-dimensional space: i, j = 1, 2, 3, ...n) (1.10)

Two important points connected with this *invariant differential line element* are:

1. Interpretation of the coefficients g_{ij} : consider a Euclidean mesh of equispaced parallelograms:

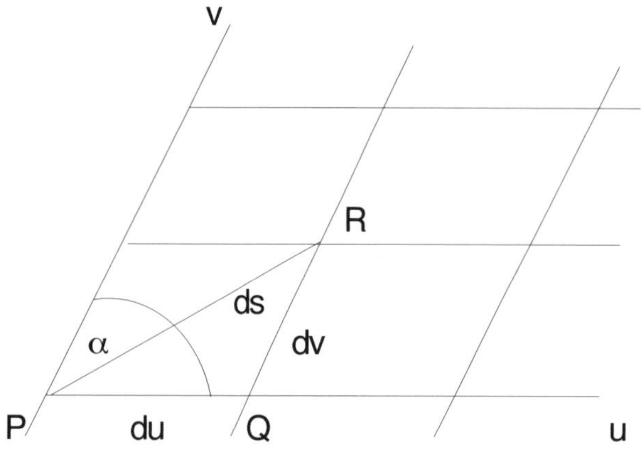

In PQR

$$ds^2 = 1.\,du^2 + 1.\,dv^2 + 2\cos\alpha\, dudv$$

$$= g_{11}du^2 + g_{22}dv^2 + 2\,g_{12}\,dudv \qquad (1.11)$$

therefore, $g_{11} = g_{22} = 1$ (the mesh-lines are equispaced)

and

$g_{12} = \cos\alpha$ where α is the angle between the u-v axes.

We see that if the mesh-lines are locally orthogonal then $g_{12} = 0$.

2. Dependence of the g_{ij}'s on the coordinate system and the local values of u, v.

A specific example will illustrate the main points of this topic: consider a point P described in three coordinate systems – Cartesian P [x, y], Polar P [r, ϕ] and Gaussian P [u, v] – and the square ds^2 of the line element in each system.

The transformation [x, y] $\rightarrow$ [r, ϕ] is

$$x = r\cos\phi \text{ and } y = r\sin\phi. \qquad (1.12\text{ a,b})$$

The transformation [r, ϕ] $\rightarrow$ [u, v] is direct, namely

$$r = u \text{ and } \phi = v.$$

Now,

$$\partial x/\partial r = \cos\phi,\ \ \partial y/\partial r = \sin\phi,\ \ \partial x/\partial \phi = -r\sin\phi,\ \ \partial y/\partial \phi = r\cos\phi$$

therefore,

$$\partial x/\partial u = \cos v,\ \ \partial y/\partial u = \sin v,\ \ \partial x/\partial v = -u\sin v,\ \ \partial y/\partial v = u\cos v.$$

The coefficients are therefore

$$g_{11} = \cos^2 v + \sin^2 v = 1, \qquad (1.13\ \text{a-c})$$

$$g_{22} = (-u\sin v)^2 + (u\cos v)^2 = u^2,$$

and

$$g_{12} = \cos(-u\sin v) + \sin v(u\cos v) = 0 \text{ (an orthogonal mesh).}$$

We therefore have

$$ds^2 = dx^2 + dy^2 \qquad (1.14\ \text{a-c})$$

$$= du^2 + u^2 dv^2$$

$$= dr^2 + r^2 d\phi^2.$$

In this example, the coefficient $g_{22} = f(u)$.

The essential point of Gaussian coordinate systems is that the coefficients g_{ij} completely characterize the surface – they are intrinsic features. We can, in principle, determine the nature of a surface by measuring the local values of the coefficients as we move over the surface. We do not need to leave a surface to study its form.

1.5 Geometry and groups

Felix Klein (1849 – 1925), introduced his influential Erlanger Program in 1872. In this program, Geometry is developed from the viewpoint of the invariants associated with *groups of transformations*. In

Euclidean Geometry, the fundamental objects are taken to be rigid bodies that remain fixed in size and shape as they are moved from place to place. The notion of a rigid body is an idealization.

Klein considered transformations of the entire plane – mappings of the set of all points in the plane onto itself. The proper set of rigid motions in the plane consists of translations and rotations. A reflection is an improper rigid motion in the plane; it is a physical impossibility in the plane itself. The set of all rigid motions – both proper and improper – forms a *group* that has the proper rigid motions as a subgroup. A group G is a set of distinct elements $\{g_i\}$ for which a *law of composition* " $\circ$ " is given such that the composition of any two elements of the set satisfies:

Closure: if g_i, g_j belong to G then $g_k = g_i \circ g_j$ belongs to G for all elements g_i, g_j,

and

Associativity: for all g_i, g_j, g_k in G, $g_i \circ (g_j \circ g_k) = (g_i \circ g_j) \circ g_k$. .

Furthermore, the set contains

A unique identity, e, such that $g_i \circ e = e \circ g_i = g_i$ for all g_i in G,

and

A unique inverse, g_i^{-1}, for every element g_i in G,

such that $g_i \circ g_i^{-1} = g_i^{-1} \circ g_i = e$.

A group that contains a finite number n of distinct elements g_n is said to be a finite group of order n.

The set of integers Z is a subset of the reals R; both sets form infinite groups under the composition of addition. Z is a "subgroup" of R.

Permutations of a set X form a group S_X under composition of functions; if a: X → X and b: X → X are permutations, the composite function ab: X → X given by ab(x) = a(b(x)) is a permutation. If the set X contains the first n positive numbers, the n! permutations form a group, the symmetric group, S_n. For example, the arrangements of the three numbers 123 form the group

$S_3 = \{ 123, 312, 231, 132, 321, 213 \}$.

If the vertices of an equilateral triangle are labeled 123, the six possible symmetry arrangements of the triangle are obtained by three successive rotations through 120° about its center of gravity, and by the three reflections in the planes I, II, III:

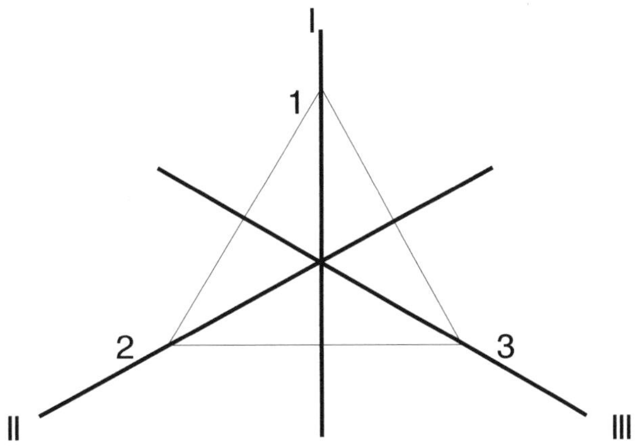

This group of "isometries" of the equilateral triangle (called the dihedral group, D_3) has the same structure as the group of permutations of three objects. The groups S_3 and D_3 are said to be isomorphic.

According to Klein, plane Euclidean Geometry is the study of those properties of plane rigid figures that are unchanged by the group of isometries. (The basic invariants are *length* and *angle*). In his development of the subject, Klein considered *Similarity Geometry* that involves isometries with a change of scale, (the basic invariant is *angle*), *Affine Geometry*, in which figures can be distorted under transformations of the form

$$x' = ax + by + c \qquad (1.15\ a,b)$$

$$y' = dx + ey + f,$$

where [x, y] are Cartesian coordinates, and a, b, c, d, e, f, are real coefficients, and *Projective Geometry*, in which all conic sections: circles, ellipses, parabolas, and hyperbolas can be transformed into one another by a projective transformation.

It will be shown that the Lorentz transformations – the fundamental transformations of events in space and time, as described by different inertial observers – form a group.

1.6 Vectors

The idea that a line with a definite length and a definite direction — a *vector* — can be used to represent a physical quantity that possesses magnitude and direction is an ancient one. The combined action of two vectors **A** and **B** is obtained by means of the parallelogram law, illustrated in the following diagram

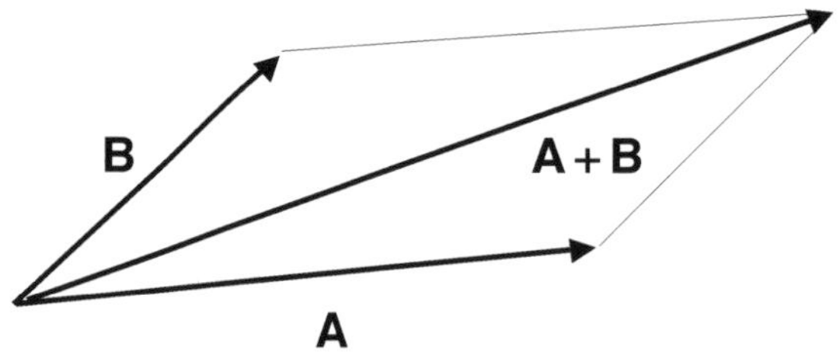

The diagonal of the parallelogram formed by **A** and **B** gives the magnitude and direction of the resultant vector **C**. Symbolically, we write

$$\mathbf{C} = \mathbf{A} + \mathbf{B} \tag{1.16}$$

in which the " = " sign has a meaning that is clearly different from its meaning in ordinary arithmetic. Galileo used this empirically - based law to obtain the resultant force acting on a body. Although a geometric approach to the study of vectors has an intuitive appeal, it will often be advantageous to use the algebraic method – particularly in the study of Einstein's Special Relativity and Maxwell's Electromagnetism.

1.7 Quaternions

In the decade 1830 - 1840, the renowned Hamilton introduced new kinds of numbers that contain *four* components, and that do not obey the commutative property of multiplication. He called the new numbers *quaternions*. A quaternion has the form

$$u + x\mathbf{i} + y\mathbf{j} + z\mathbf{k} \tag{1.17}$$

in which the quantities **i**, **j**, **k** are akin to the quantity $i = \sqrt{-1}$ in complex numbers, $x + iy$. The component u forms the scalar part, and the three components $x\mathbf{i} + y\mathbf{j} + z\mathbf{k}$ form the vector part of the quaternion. The coefficients x, y, z can be considered to be the Cartesian components of a point P in space. The quantities **i**, **j**, **k** are qualitative units that are directed along the coordinate axes. Two quaternions are equal if their scalar parts are equal, and if their coefficients x, y, z of **i**, **j**, **k** are respectively equal. The sum of two quaternions is a quaternion. In operations that involve quaternions, the usual rules of multiplication hold except in those terms in which products of **i**, **j**, **k** occur — in these terms, the commutative law does not hold. For example

$$\mathbf{j}\,\mathbf{k} = \mathbf{i},\ \mathbf{k}\,\mathbf{j} = -\mathbf{i},\ \mathbf{k}\,\mathbf{i} = \mathbf{j},\ \mathbf{i}\,\mathbf{k} = -\mathbf{j},\ \mathbf{i}\,\mathbf{j} = \mathbf{k},\ \mathbf{j}\,\mathbf{i} = -\mathbf{k}, \tag{1.18}$$

(these products obey a right-hand rule),

and

$$\mathbf{i}^2 = \mathbf{j}^2 = \mathbf{k}^2 = -1. \text{ (Note the relation to } i^2 = -1\text{).} \tag{1.19}$$

The product of two quaternions does not commute. For example, if

$$p = 1 + 2\mathbf{i} + 3\mathbf{j} + 4\mathbf{k}, \text{ and } q = 2 + 3\mathbf{i} + 4\mathbf{j} + 5\mathbf{k}$$

then

$$pq = -36 + 6\mathbf{i} + 12\mathbf{j} + 12\mathbf{k}$$

whereas

$$qp = -36 + 23\mathbf{i} - 2\mathbf{j} + 9\mathbf{k}.$$

Multiplication is associative.

Quaternions can be used as operators to rotate and scale a given vector into a new vector:

$$(a + b\mathbf{i} + c\mathbf{j} + d\mathbf{k})(x\mathbf{i} + y\mathbf{j} + z\mathbf{k}) = (x'\mathbf{i} + y'\mathbf{j} + z'\mathbf{k})$$

If the law of composition is quaternionic multiplication then the set

$$Q = \{\pm 1, \pm\mathbf{i}, \pm\mathbf{j}, \pm\mathbf{k}\}$$

is found to be a group of order 8. It is a non-commutative group.

Hamilton developed the Calculus of Quaternions. He considered, for example, the properties of the differential operator:

$$\nabla = \mathbf{i}(\partial/\partial x) + \mathbf{j}(\partial/\partial y) + \mathbf{k}(\partial/\partial z). \qquad (1.20)$$

(He called this operator "nabla").

If f(x, y, z) is a scalar point function (single-valued) then

$$\nabla f = \mathbf{i}(\partial f/\partial x) + \mathbf{j}(\partial f/\partial y) + \mathbf{k}(\partial f/\partial z)\text{, a vector.}$$

If

$$\mathbf{v} = v_1\mathbf{i} + v_2\mathbf{j} + v_3\mathbf{k}$$

is a continuous vector point function, where the v_i's are functions of x, y, and z, Hamilton introduced the operation

$$\nabla\mathbf{v} = (\mathbf{i}\partial/\partial x + \mathbf{j}\partial/\partial y + \mathbf{k}\partial/\partial z)(v_1\mathbf{i} + v_2\mathbf{j} + v_3\mathbf{k}) \qquad (1.21)$$

$$= -(\partial v_1/\partial x + \partial v_2/\partial y + \partial v_3/\partial z)$$

$$+ (\partial v_3/\partial y - \partial v_2/\partial z)\mathbf{i} + (\partial v_1/\partial z - \partial v_3/\partial x)\mathbf{j} + (\partial v_2/\partial x - \partial v_1/\partial y)\mathbf{k}$$

$$= \text{a quaternion.}$$

The scalar part is the negative of the "divergence of **v**" (a term due to Clifford), and the vector part is the "curl of **v**" (a term due to Maxwell). Maxwell used the repeated operator ∇^2, which he called the Laplacian.

1.8 3 – Vector Analysis

Gibbs, in his notes for Yale students, written in the period 1881 - 1884, and Heaviside, in articles published in the *Electrician* in the 1880's, independently developed 3-dimensional Vector Analysis as a subject in its own right — detached from quaternions.

In the Sciences, and in parts of Mathematics (most notably in Analytical and Differential Geometry), their methods are widely used. Two kinds of vector multiplication were introduced: scalar multiplication and vector multiplication. Consider two vectors **v** and **v**´ where

$$\mathbf{v} = v_1\mathbf{e}_1 + v_2\mathbf{e}_2 + v_3\mathbf{e}_3$$

and

$$\mathbf{v}' = v_1'\mathbf{e}_1 + v_2'\mathbf{e}_2 + v_3'\mathbf{e}_3 .$$

The quantities $\mathbf{e}_1$, $\mathbf{e}_2$, and $\mathbf{e}_3$ are vectors of unit length pointing along mutually orthogonal axes, labeled 1, 2, and 3.

i) The scalar multiplication of **v** and **v**´ is defined as

$$\mathbf{v} \cdot \mathbf{v}' = v_1v_1' + v_2v_2' + v_3v_3', \qquad (1.22)$$

where the unit vectors have the properties

$$\mathbf{e}_1 \cdot \mathbf{e}_1 = \mathbf{e}_2 \cdot \mathbf{e}_2 = \mathbf{e}_3 \cdot \mathbf{e}_3 = 1, \text{ and} \qquad (1.23)$$

$$\mathbf{e}_1 \cdot \mathbf{e}_2 = \mathbf{e}_2 \cdot \mathbf{e}_1 = \mathbf{e}_1 \cdot \mathbf{e}_3 = \mathbf{e}_3 \cdot \mathbf{e}_1 = \mathbf{e}_2 \cdot \mathbf{e}_3 = \mathbf{e}_3 \cdot \mathbf{e}_2 = 0. \qquad (1.24)$$

The most important property of the scalar product of two vectors is its invariance under rotations and translations of the coordinates. (See Chapter **1**).

ii) The vector product of two vectors **v** and **v**´ is defined as

$$\mathbf{v} \times \mathbf{v}' = \begin{vmatrix} \mathbf{e}_1 & \mathbf{e}_2 & \mathbf{e}_3 \\ v_1 & v_2 & v_3 \\ v_1' & v_2' & v_3' \end{vmatrix} \quad \text{(where |. . . |is the determinant)} \qquad (1.25)$$

$$= (v_2 v_3' - v_3 v_2')\mathbf{e}_1 + (v_3 v_1' - v_1 v_3')\mathbf{e}_2 + (v_1 v_2' - v_2 v_1')\mathbf{e}_3 .$$

The unit vectors have the properties

$$\mathbf{e}_1 \times \mathbf{e}_1 = \mathbf{e}_2 \times \mathbf{e}_2 = \mathbf{e}_3 \times \mathbf{e}_3 = 0 \qquad (1.26\text{ a,b})$$

(note that these properties differ from the quaternionic products of the **i**, **j**, **k**'s),

and

$$\mathbf{e}_1 \times \mathbf{e}_2 = \mathbf{e}_3 \,,\ \mathbf{e}_2 \times \mathbf{e}_1 = -\mathbf{e}_3 \,,\ \mathbf{e}_2 \times \mathbf{e}_3 = \mathbf{e}_1 \,,\ \mathbf{e}_3 \times \mathbf{e}_2 = -\mathbf{e}_1 \,,\ \mathbf{e}_3 \times \mathbf{e}_1 = \mathbf{e}_2 \,,\ \mathbf{e}_1 \times \mathbf{e}_3 = -\mathbf{e}_2$$

These non-commuting vectors, or "cross products" obey the standard right-hand-rule.

The vector product of two parallel vectors is zero even when neither vector is zero.

The non-associative property of a vector product is illustrated in the following example

$$\mathbf{e}_1 \times \mathbf{e}_2 \times \mathbf{e}_2 = (\mathbf{e}_1 \times \mathbf{e}_2) \times \mathbf{e}_2 = \mathbf{e}_3 \times \mathbf{e}_2 = -\mathbf{e}_1$$

$$= \mathbf{e}_1 \times (\mathbf{e}_2 \times \mathbf{e}_2) = 0.$$

Important operations in Vector Analysis that follow directly from those introduced in the theory of quaternions are:

1) the *gradient* of a scalar function $f(x_1, x_2, x_3)$

$$\nabla f = (\partial f/\partial x_1)\mathbf{e}_1 + (\partial f/\partial x_2)\mathbf{e}_2 + (\partial f/\partial x_3)\mathbf{e}_3 , \qquad (1.27)$$

2) the *divergence* of a vector function **v**

$$\nabla \cdot \mathbf{v} = \partial v_1/\partial x_1 + \partial v_2/\partial x_2 + \partial v_3/\partial x_3 \qquad (1.28)$$

where **v** has components v_1, v_2, v_3 that are functions of x_1, x_2, x_3 , and

3) the *curl* of a vector function **v**

$$\nabla \times \mathbf{v} = \begin{vmatrix} \mathbf{e}_1 & \mathbf{e}_2 & \mathbf{e}_3 \\ \partial/\partial x_1 & \partial/\partial x_2 & \partial/\partial x_3 \\ v_1 & v_2 & v_3 \end{vmatrix} . \qquad (1.29)$$

The physical significance of these operations is discussed later.

1.9 Linear algebra and n-vectors

A major part of Linear Algebra is concerned with the extension of the algebraic properties of vectors in the plane (2-vectors), and in space (3-vectors), to vectors in higher dimensions (n-vectors). This area of study has its origin in the work of Grassmann (1809 - 77), who generalized the quaternions (4-component hyper-complex numbers), introduced by Hamilton.

An n-dimensional vector is defined as an ordered column of numbers

$$\mathbf{x}_n = \begin{pmatrix} x_1 \\ x_2 \\ . \\ . \\ x_n \end{pmatrix} \quad (1.30)$$

It will be convenient to write this as an ordered row in square brackets

$$\mathbf{x}_n = [x_1, x_2, \ldots x_n]\,. \quad (1.31)$$

The transpose of the column vector is the row vector

$$\mathbf{x}_n^T = (x_1, x_2, \ldots x_n). \quad (1.32)$$

The numbers $x_1, x_2, \ldots x_n$ are called the components of **x**, and the integer n is the dimension of **x**. The order of the components is important, for example

$$[1, 2, 3] \neq [2, 3, 1].$$

The two vectors $\mathbf{x} = [x_1, x_2, \ldots x_n]$ and $\mathbf{y} = [y_1, y_2, \ldots y_n]$ are equal if

$$x_i = y_i \ (i = 1 \text{ to } n).$$

The laws of Vector Algebra are

1. $\mathbf{x} + \mathbf{y} = \mathbf{y} + \mathbf{x}$. (1.33 a-e)

2. $[\mathbf{x} + \mathbf{y}] + \mathbf{z} = \mathbf{x} + [\mathbf{y} + \mathbf{z}]$.

3. $a[\mathbf{x} + \mathbf{y}] = a\mathbf{x} + a\mathbf{y}$ where a is a scalar .

4. $(a + b)\mathbf{x} = a\mathbf{x} + b\mathbf{y}$ where a,b are scalars .

5. $(ab)\mathbf{x} = a(b\mathbf{x})$ where a,b are scalars .

If $a = 1$ and $b = -1$ then

$\mathbf{x} + [-\mathbf{x}] = \mathbf{0}$,

where $\mathbf{0} = [0, 0, \dots 0]$ is the zero vector.

The vectors $\mathbf{x} = [x_1, x_2, \dots x_n]$ and $\mathbf{y} = [y_1, y_2 \dots y_n]$ can be added to give their sum or resultant:

$$\mathbf{x} + \mathbf{y} = [x_1 + y_1, x_2 + y_2, \dots, x_n + y_n]. \qquad (1.34)$$

The set of vectors that obeys the above rules is called the *space of all n-vectors* or the *vector space of dimension n.*

In general, a vector $\mathbf{v} = a\mathbf{x} + b\mathbf{y}$ lies in the plane of **x** and **y**. The vector **v** is said to depend linearly on **x** and **y** — it is a linear combination of **x** and **y**.

A k-vector **v** is said to depend linearly on the vectors $\mathbf{u}_1, \mathbf{u}_2, \dots \mathbf{u}_k$ if there are scalars a_i such that

$$\mathbf{v} = a_1\mathbf{u}_1 + a_2\mathbf{u}_2 + \dots a_k\mathbf{u}_k . \qquad (1.35)$$

For example

[3, 5, 7] = [3, 6, 6] + [0, –1, 1] = 3[1, 2, 2] + 1[0, –1, 1], a linear combination of the vectors [1, 2, 2] and [0, –1, 1].

A set of vectors $\mathbf{u}_1, \mathbf{u}_2, \dots\mathbf{u}_k$ is called *linearly dependent* if one of these vectors depends linearly on the rest. For example, if

$\mathbf{u}_1 = a_2\mathbf{u}_2 + a_3\mathbf{u}_3 + \dots + a_k\mathbf{u}_k$ the set $\mathbf{u}_1, \dots\mathbf{u}_k$ is linearly dependent. (1.36)

If none of the vectors $\mathbf{u}_1, \mathbf{u}_2, \dots\mathbf{u}_k$ can be written linearly in terms of the remaining ones we say that the vectors are *linearly independent.*

Alternatively, the vectors $\mathbf{u}_1, \mathbf{u}_2, \dots\mathbf{u}_k$ are linearly dependent if and only if there is an equation of the form

$$c_1\mathbf{u}_1 + c_2\mathbf{u}_2 + \ldots c_k\mathbf{u}_k = \mathbf{0}, \tag{1.37}$$

in which the scalars c_i are not all zero.

Consider the vectors $\mathbf{e}_i$ obtained by putting the ith-component equal to 1, and all the other components equal to zero:

$\mathbf{e}_1 = [1, 0, 0, \ldots 0]$

$\mathbf{e}_2 = [0, 1, 0, \ldots 0]$

$\ldots$

then every vector of dimension n depends linearly on $\mathbf{e}_1, \mathbf{e}_2, \ldots \mathbf{e}_n$, thus

$$\mathbf{x} = [x_1, x_2, \ldots x_n]$$

$$= x_1\mathbf{e}_1 + x_2\mathbf{e}_2 + \ldots x_n\mathbf{e}_n. \tag{1.38}$$

The $\mathbf{e}_i$'s are said to *span the space* of all n-vectors; they form a *basis*. Every basis of an n-space has exactly n elements. The connection between a vector **x** and a definite coordinate system is made by choosing a set of basis vectors $\mathbf{e}_i$.

1.10 The geometry of vectors

The laws of vector algebra can be interpreted geometrically for vectors of dimension 2 and 3. Let the zero vector represent the origin of a coordinate system, and let the 2-vectors, **x** and **y**, correspond to points in the plane: P $[x_1, x_2]$ and Q $[y_1, y_2]$. The vector sum $\mathbf{x} + \mathbf{y}$ is represented by the point R, as show

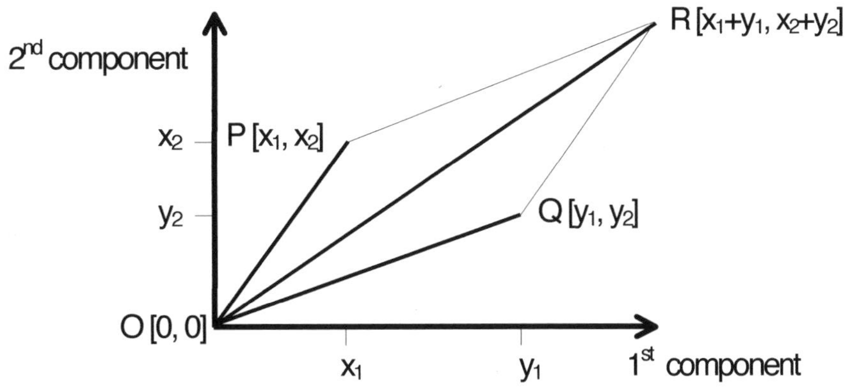

R is in the plane OPQ, even if **x** and **y** are 3-vectors.

Every vector point on the line OR represents the sum of the two corresponding vector points on the lines OP and OQ. We therefore introduce the concept of the directed vector lines $\overrightarrow{OP}$, $\overrightarrow{OQ}$, and $\overrightarrow{OR}$, related by the vector equation

$$\overrightarrow{OP} + \overrightarrow{OQ} = \overrightarrow{OR}. \tag{1.39}$$

A vector **V** can be represented as a line of length OP pointing in the direction of the unit vector **v**, thus

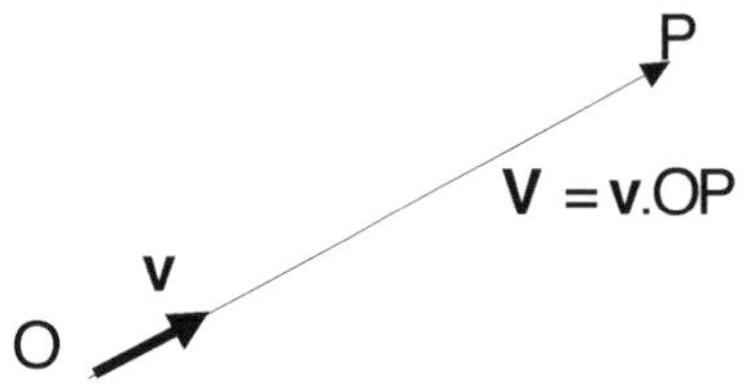

A vector **V** is unchanged by a pure displacement:

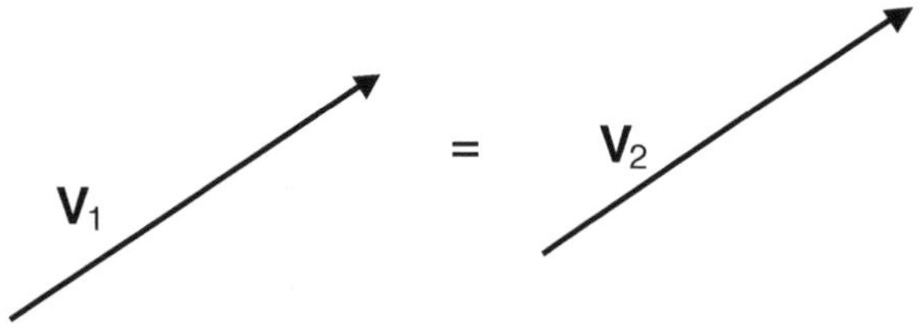

where the " = " sign means equality in magnitude and direction.

Two classes of vectors will be met in future discussions; they are

1. *Polar vectors*: the vector is drawn in the direction of the physical quantity being represented, for example a velocity, and

2. *Axial vectors*: the vector is drawn parallel to the axis about which the physical quantity acts, for example an angular velocity.

The associative property of the sum of vectors can be readily demonstrated, geometrically

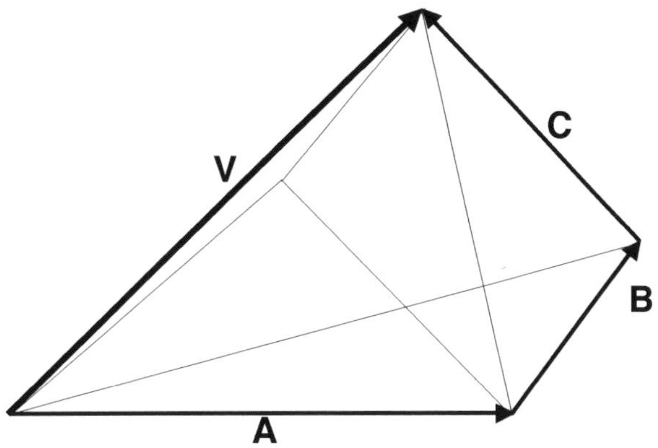

We see that

$$\mathbf{V} = \mathbf{A} + \mathbf{B} + \mathbf{C} = (\mathbf{A} + \mathbf{B}) + \mathbf{C} = \mathbf{A} + (\mathbf{B} + \mathbf{C}) = (\mathbf{A} + \mathbf{C}) + \mathbf{B} . \tag{1.40}$$

The process of vector addition can be reversed; a vector **V** can be decomposed into the sum of n vectors of which (n – 1) are arbitrary, and the nth vector closes the polygon. The vectors need not be in the same plane. A special case of this process is the decomposition of a 3-vector into its Cartesian components.

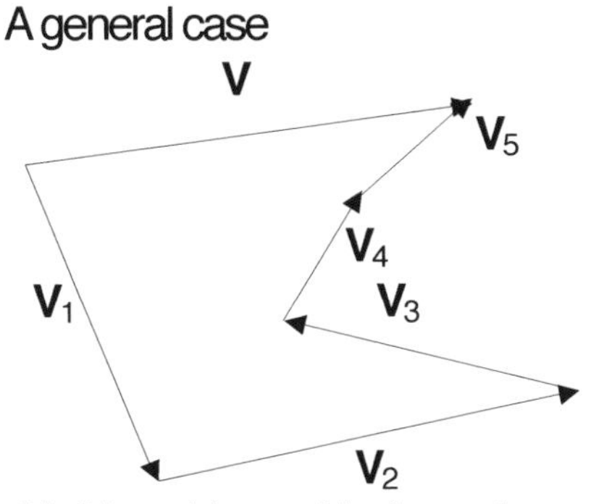

$\mathbf{V}_1, \mathbf{V}_2, \mathbf{V}_3, \mathbf{V}_4$: arbitrary , $\mathbf{V}_5$ closes the polygon

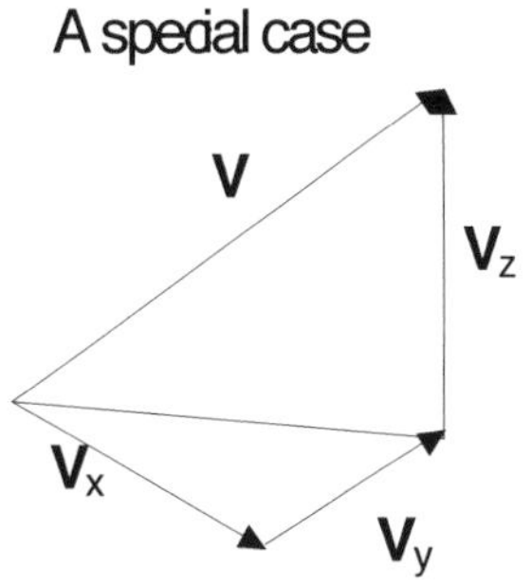

$\mathbf{V}_z$ closes the polygon

The *vector product* of **A** and **B** is an axial vector, perpendicular to the plane containing **A** and **B.**

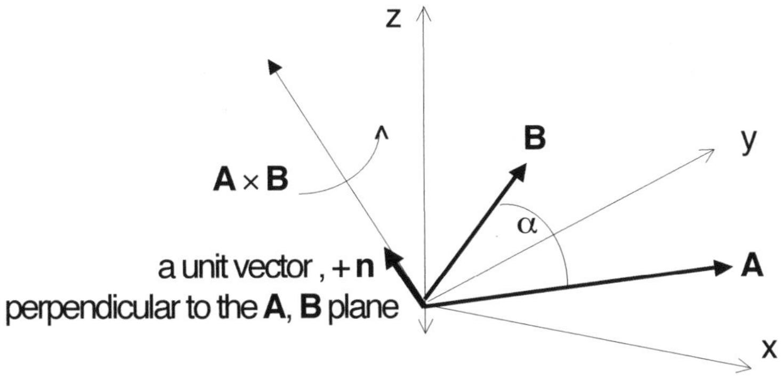

$$\mathbf{A} \times \mathbf{B} = AB \sin\alpha\, \mathbf{n} = -\mathbf{B} \times \mathbf{A} \tag{1.41}$$

1.11 Linear Operators and Matrices

Transformations from a coordinate system [x, y] to another system [x′, y′], without shift of the origin, or from a point P [x, y] to another point P′ [x′, y′], in the same system, with the form

$$x' = ax + by$$

$$y' = cx + dy$$

where a, b, c, d are real coefficients, can be written in matrix notation, as follows

$$\begin{pmatrix} x' \\ y' \end{pmatrix} = \begin{pmatrix} a & b \\ c & d \end{pmatrix} \begin{pmatrix} x \\ y \end{pmatrix}.$$

Symbolically,

$$\mathbf{x}' = \mathbf{Mx},\ \mathbf{x} = [x, y],\ \text{and } \mathbf{x}' = [x', y'],\ \text{both } \textit{column}\ \text{2-vectors,} \quad (1.42)$$

and

$$\mathbf{M} = \begin{pmatrix} a & b \\ c & d \end{pmatrix},$$

a 2×2 matrix operator that changes [x, y] into [x′, y′].

In general, **M** transforms a unit square into a parallelogram:

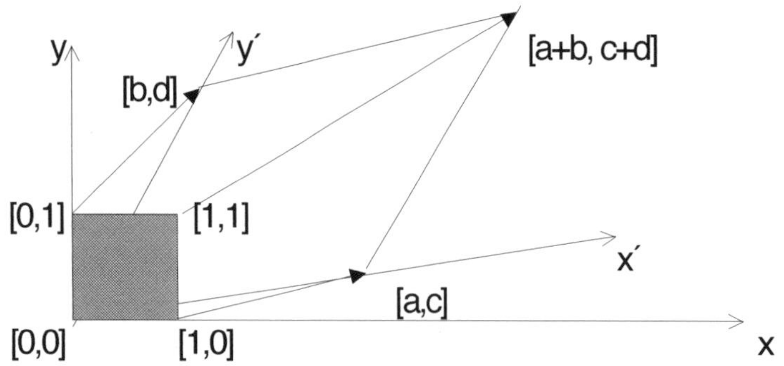

This transformation plays a key rôle in Einstein's Special Theory of Relativity (see later discussion).

1.12 Rotation operators

Consider the rotation of an x, y coordinate system about the origin through an angle ϕ:

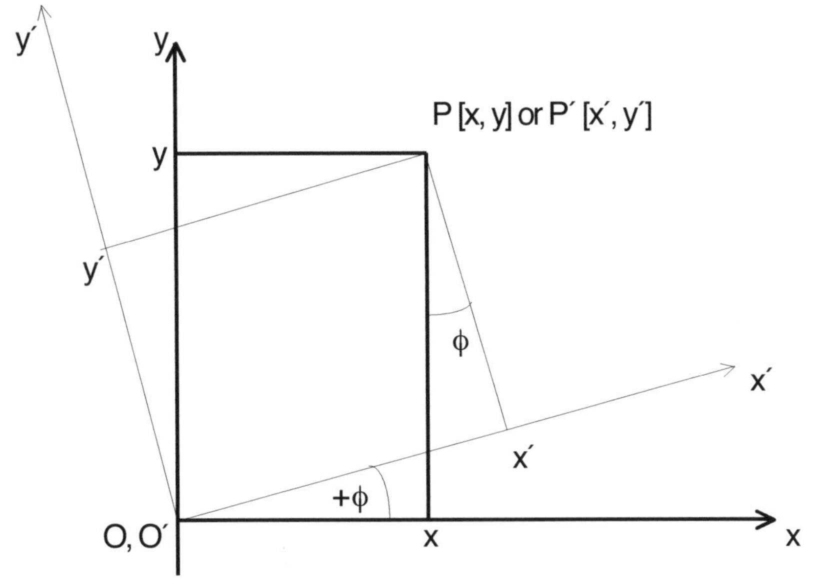

From the diagram, we see that

$$x' = x\cos\phi + y\sin\phi \text{ and } y' = -x\sin\phi + y\cos\phi$$

or

$$\begin{pmatrix} x' \\ y' \end{pmatrix} = \begin{pmatrix} \cos\phi & \sin\phi \\ -\sin\phi & \cos\phi \end{pmatrix} \begin{pmatrix} x \\ y \end{pmatrix}.$$

Symbolically,

$$\mathbf{P}' = \Re_c(\phi)\mathbf{P} \qquad (1.43)$$

where

$$\Re_c(\phi) = \begin{pmatrix} \cos\phi & \sin\phi \\ -\sin\phi & \cos\phi \end{pmatrix} \text{ is the } \textit{rotation operator}.$$

The subscript c denotes a rotation of the *coordinates* through an angle $+\phi$.

The inverse operator, $\Re_c^{-1}(\phi)$, is obtained by reversing the angle of rotation: $+\phi \to -\phi$.

We see that matrix product

$$\Re_c^{-1}(\phi)\Re_c(\phi) = \Re_c^{T}(\phi)\Re_c(\phi) = \mathbf{I} \qquad (1.44)$$

where the superscript T indicates the transpose (rows $\Leftrightarrow$ columns), and

$$\mathbf{I} = \begin{pmatrix} 1 & 0 \\ 0 & 1 \end{pmatrix} \text{ is the identity operator.} \qquad (1.45)$$

Eq.(1.44) is the defining property of an *orthogonal matrix.*

If we leave the axes fixed and rotate the point P[x, y] to P′[x′, y′], then we have

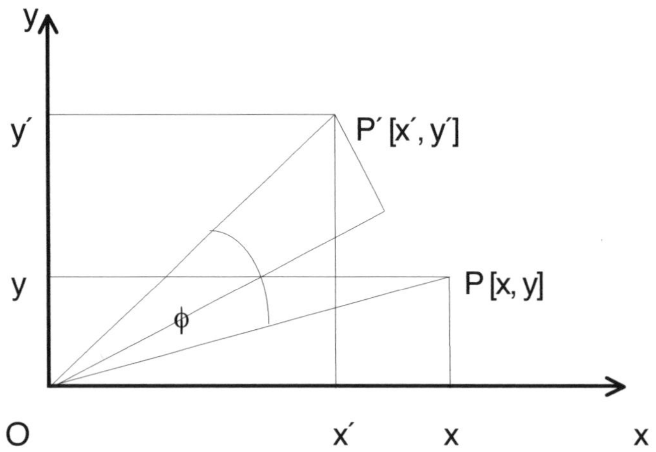

From the diagram, we see that

$$x' = x\cos\phi - y\sin\phi, \text{ and } y' = x\sin\phi + y\cos\phi$$

or

$$\mathbf{P}' = \Re_V(\phi)\mathbf{P} \qquad (1.46)$$

where

$$\Re_V(\phi) = \begin{pmatrix} \cos\phi & -\sin\phi \\ \sin\phi & \cos\phi \end{pmatrix}, \text{ the operator that rotates a vector through } +\phi.$$

1.13 Components of a vector under coordinate rotations

Consider a vector **V** [$\mathbf{v}_x$, $\mathbf{v}_y$], and the same vector **V**′ with components [$\mathbf{v}_{x'}$, $\mathbf{v}_{y'}$], in a coordinate system (primed), rotated through an angle $+\phi$.

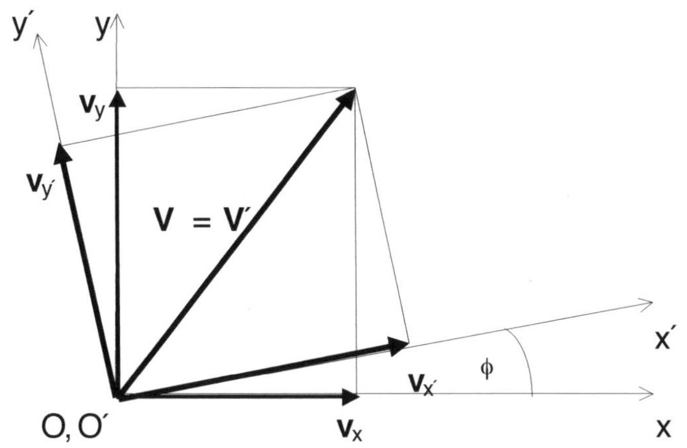

We have met the transformation [x, y] → [x´, y´] under the operation $\Re_c(\phi)$; here, we have the same transformation but now it operates on the components of the vector, $\mathbf{v}_x$ and $\mathbf{v}_y$,

$$[\mathbf{v}_{x´}, \mathbf{v}_{y´}] = \Re_c(\phi)[\mathbf{v}_x, \mathbf{v}_y]. \qquad (1.47)$$

PROBLEMS

1-1 i) If $u = 3^{x/y}$ show that $\partial u/\partial x = (3^{x/y}\ln 3)/y$ and $\partial u/\partial y = (-3^{x/y}x\ln 3)/y^2$.

ii) If $u = \ln\{(x^3 + y)/x^2\}$ show that $\partial u/\partial x = (x^3 - 2y)/(x(x^3 + y))$ and $\partial u/\partial y = 1/(x^3 + y)$.

1-2 Calculate the second partial derivatives of

$$f(x, y) = (1/\sqrt{y})\exp\{-(x - a)^2/4y\},\ a = \text{constant}.$$

1-3 Check the answers obtained in problem 1-2 by showing that the function f(x, y) in 1-2 is a solution of the partial differential equation $\partial^2 f/\partial x^2 - \partial f/\partial y = 0$.

1-4 If $f(x, y, z) = 1/(x^2 + y^2 + z^2)^{1/2} = 1/r$, show that $f(x, y, z) = 1/r$ is a solution of Laplace's equation

$$\partial^2 f/\partial x^2 + \partial^2 f/\partial y^2 + \partial^2 f/\partial z^2 = 0.$$

This important equation occurs in many branches of Physics.

1-5 At a given instant, the radius of a cylinder is r(t) = 4cm and its height is h(t) = 10cm. If r(t) and h(t) are both changing at a rate of 2 cm.s^{-1}, show that the instantaneous increase in the volume of the cylinder is $192\pi\,cm^3.s^{-1}$.

1-6 The transformation between Cartesian coordinates [x, y, z] and spherical polar coordinates [r, θ, φ] is

$$x = r\sin\theta\cos\phi,\ y = r\sin\theta\sin\phi,\ z = r\cos\theta.$$

Show, by calculating all necessary partial derivatives, that the square of the line element is

$$ds^2 = dr^2 + r^2\sin^2\theta\, d\phi^2 + r^2 d\theta^2.$$

Obtain this result using geometrical arguments. This form of the square of the line element will be used on several occasions in the future.

1-7 Prove that the inverse of each element of a group is unique.

1-8 Prove that the set of positive rational numbers does not form a group under division.

1-9 A finite group of order n has n^2 products that may be written in an n×n array, called the group multiplication table. For example, the 4th-roots of unity {e, a, b, c} = {±1, ±i}, where i = $\sqrt{-1}$, forms a group under multiplication ($1i = i$, $i(-i) = 1$, $i^2 = -1$, $(-i)^2 = -1$, etc.) with a multiplication table

	e = 1	a = i	b = −1	c = −i
e	1	i	−1	−i
a	i	−1	−i	1
b	−1	−i	1	i
c	−i	1	i	−1

In this case, the table is symmetric about the main diagonal; this is a characteristic feature of a group in which all products commute (ab = ba) — it is an Abelian group.

If G is the dihedral group D_3, discussed in the text, where $G = \{e, a, a^2, b, c, d\}$, where e is the identity, obtain the group multiplication table. Is it an Abelian group?. Notice that the three elements $\{e, a, a^2\}$ form a subgroup of G, whereas the three elements {b, c, d} do not; there is no identity in this subset.

The group D_3 has the same multiplication table as the group of permutations of three objects. This is the condition that signifies group isomorphism.

1-10 Are the sets

i) {[0, 1, 1], [1, 0, 1], [1, 1, 0]}

and

ii) {[1, 3, 5, 7], [4, −3, 2, 1], [2, 1, 4, 5]}

linearly dependent? Explain.

1-11 i) Prove that the vectors [0, 1, 1], [1, 0, 1], [1, 1, 0] form a basis for Euclidean space R^3.

ii) Do the vectors [1, i] and [i, −1], $(i = \sqrt{-1})$, form a basis for the complex space C^2?

1-12 Interpret the linear independence of two 3-vectors geometrically.

1-13 i) If $\mathbf{X} = [1, 2, 3]$ and $\mathbf{Y} = [3, 2, 1]$, prove that their cross product is orthogonal to the **X**-**Y** plane.

ii) If **X** and **Y** are 3-vectors, prove that $\mathbf{X}\times\mathbf{Y} = 0$ iff **X** and **Y** are linearly dependent.

1-14 If

$$\mathbf{T} = \begin{pmatrix} a_{11} & a_{12} & a_{13} \\ a_{21} & a_{22} & a_{23} \\ 0 & 0 & 1 \end{pmatrix}$$

represents a linear transformation of the plane under which *distance* is an *invariant*, show that the following relations must hold :

$$a_{11}^2 + a_{21}^2 = a_{12}^2 + a_{22}^2 = 1, \text{ and } a_{11}a_{12} + a_{21}a_{22} = 0.$$

2

KINEMATICS: THE GEOMETRY OF MOTION

2.1 Velocity and acceleration

The most important concepts in Kinematics — a subject in which the properties of the forces responsible for the motion are ignored — can be introduced by studying the simplest of all motions, namely that of a point P moving in a straight line.

Let a point P [t, x] be at a distance x from a fixed point O at a time t, and let it be at a point P′ [t′, x′] = P′[t + Δt, x + Δx] at a time Δt later. The *average speed* of P in the interval Δt is

$$\langle v_p \rangle = \Delta x/\Delta t. \tag{2.1}$$

If the ratio Δx/Δt is not constant in time, we define the *instantaneous speed* of P at time t as the limiting value of the ratio as Δt → 0:

$$v_p = v_p(t) = \text{limit as } \Delta t \to 0 \text{ of } \Delta x/\Delta t = dx/dt = \dot{x} = v_x .$$

The instantaneous speed is the *magnitude of a vector* called the *instantaneous velocity* of P:

$$\mathbf{v} = d\mathbf{x}/dt \text{, a quantity that has both magnitude and direction.} \tag{2.2}$$

A *space-time curve* is obtained by plotting the positions of P as a function of t:

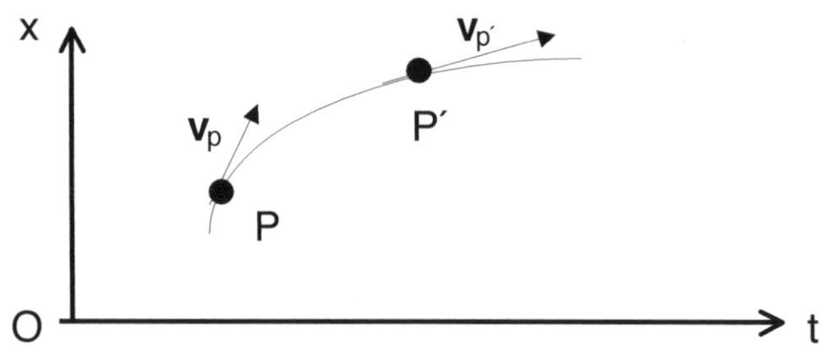

The tangent of the angle made by the tangent to the curve at any point gives the value of the instantaneous speed at the point.

The *instantaneous acceleration*, **a** , of the point P is given by the time rate-of-change of the velocity

$$\mathbf{a} = d\mathbf{v}/dt = d(d\mathbf{x}/dt)/dt = d^2\mathbf{x}/dt^2 = \ddot{\mathbf{x}}. \qquad (2.3)$$

A change of variable from t to x gives

$$a = dv/dt = dv(dx/dt)/dx = v(dv/dx). \qquad (2.4)$$

This is a useful relation when dealing with problems in which the velocity is given as a function of the position. For example

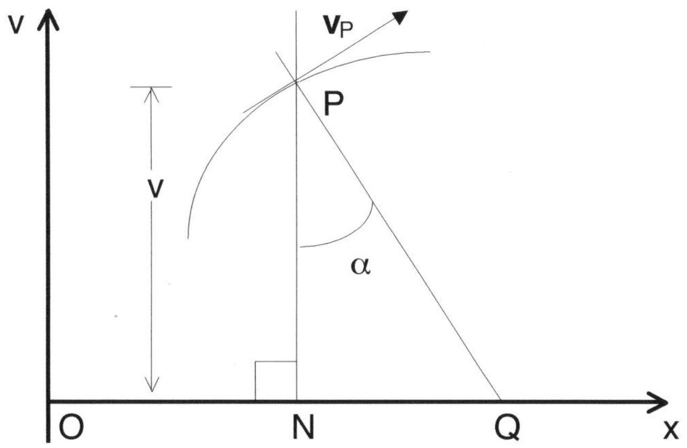

The gradient is dv/dx and $\tan\alpha = dv/dx$, therefore

NQ, the *subnormal*, $= v(dv/dx) = a_p$, the acceleration of P. (2.5)

The area under a curve of the *speed as a function of time* between the times t_1 and t_2 is

$$[A]_{[t1,,t2]} = \int_{[t1,t2]} v(t)dt = \int_{[t1,t2]} (dx/dt)dt = \int_{[x1,x2]} dx = (x_2 - x_1)$$

$$= \text{distance traveled in the time } t_2 - t_1. \qquad (2.6)$$

The solution of a kinematical problem is sometimes simplified by using a graphical method, for example:

A point A moves along an x-axis with a constant speed v_A. Let it be at the origin O ($x = 0$) at time $t = 0$. It continues for a distance x_A, at which point it decelerates at a constant rate, finally stopping at a distance X from O at time T. A second point B moves away from O in the +x-direction with constant acceleration. Let it begin its motion at $t = 0$. It continues to accelerate until it reaches a maximum speed v_B^{max} at a time t_B^{max} when

at x_B^{max} from O. At x_B^{max}, it begins to decelerate at a constant rate, finally stopping at X at time T: To prove that the maximum speed of B during its motion is

$v_B^{max} = v_A\{1 - (x_A/2X)\}^{-1}$, a value that is independent of the time at which the maximum speed is reached.

The velocity-time curves of the points are

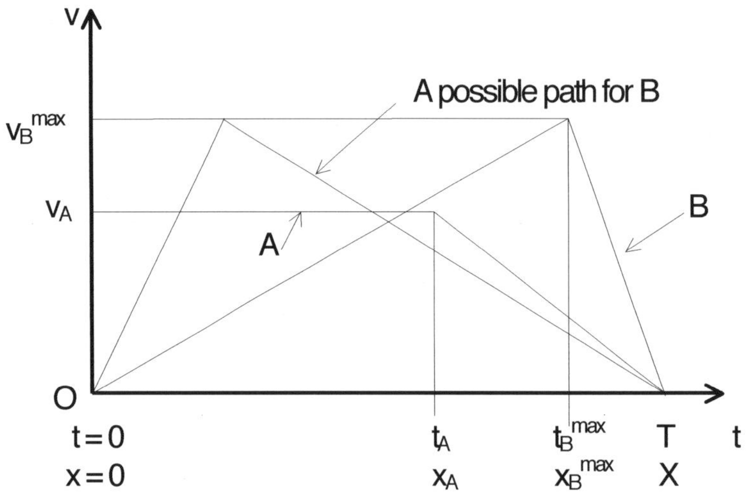

The areas under the curves give $X = v_At_A + v_A(T - t_A)/2 = v_B^{max}T/2$, so that

$v_B^{max} = v_A(1 + (t_A/T))$, but $v_AT = 2X - x_A$, therefore $v_B^{max} = v_A\{1 - (x_A/2X)\}^{-1} \neq f(t_B^{max})$.

2.2 Differential equations of kinematics

If the acceleration is a known function of time then the differential equation

$$a(t) = dv/dt \tag{2.7}$$

can be solved by performing the integrations (either analytically or numerically)

$$\int a(t)dt = \int dv \tag{2.8}$$

If a(t) is constant then the result is simply

$at + C = v$, where C is a constant that is given by the initial conditions.

Let $v = u$ when $t = 0$ then $C = u$ and we have

$$at + u = v. \tag{2.9}$$

This is the standard result for motion under constant acceleration.

We can continue this approach by writing:

$$v = dx/dt = u + at.$$

Separating the variables,

$$dx = udt + atdt.$$

Integrating gives

$$x = ut + (1/2)at^2 + C' \quad \text{(for constant a).}$$

If $x = 0$ when $t = 0$ then $C' = 0$, and

$$x(t) = ut + (1/2)at^2. \tag{2.10}$$

Multiplying this equation throughout by 2a gives

$$2ax = 2aut + (at)^2$$

$$= 2aut + (v - u)^2$$

and therefore, rearranging, we obtain

$$v^2 = 2ax - 2aut + 2vu - u^2$$

$$= 2ax + 2u(v - at) - u^2$$

$$= 2ax + u^2. \tag{2.11}$$

In general, the acceleration is a given function of time or distance or velocity:

1) If $a = f(t)$ then

$$a = dv/dt = f(t), \tag{2.12}$$

$$dv = f(t)dt,$$

therefore

$$v = \int f(t)dt + C (\text{a constant}).$$

This equation can be written

$$v = dx/dt = F(t) + C,$$

therefore

$$dx = F(t)dt + C\,dt.$$

Integrating gives

$$x(t) = \int F(t)dt + C\,t + C'. \qquad (2.13)$$

The constants of integration can be determined if the velocity and the position are known at a given time.

2) If $a = g(x) = v(dv/dx)$ then (2.14)

$$vdv = g(x)dx.$$

Integrating gives

$$v^2 = 2\int g(x)dx + D,$$

therefore

$$v^2 = G(x) + D$$

so that

$$v = (dx/dt) = \pm\sqrt{(G(x) + D)}. \qquad (2.15)$$

Integrating this equation leads to

$$\pm\int dx/\{\sqrt{(G(x) + D)}\} = t + D'. \qquad (2.16)$$

Alternatively, if

$$a = d^2x/dt^2 = g(x)$$

then, multiplying throughout by $2(dx/dt)$ gives

$$2(dx/dt)(d^2x/dt^2) = 2(dx/dt)g(x).$$

Integrating then gives

$$(dx/dt)^2 = 2\int g(x)dx + D \text{ etc.}$$

As an example of this method, consider the equation of simple harmonic motion (see later discussion)

$$d^2x/dt^2 = -\omega^2 x. \tag{2.17}$$

Multiply throughout by 2(dx/dt), then

$$2(dx/dt)d^2x/dt^2 = -2\omega^2 x(dx/dt).$$

This can be integrated to give

$$(dx/dt)^2 = -\omega^2 x^2 + D.$$

If $dx/dt = 0$ when $x = A$ then $D = \omega^2 A^2$, therefore

$$(dx/dt)^2 = \omega^2(A^2 - x^2) = v^2,$$

so that

$$dx/dt = \pm\omega\surd(A^2 - x^2).$$

Separating the variables, we obtain

$$-dx/\{\surd(A^2 - x^2)\} = \omega dt.$$ (The minus sign is chosen because dx and dt have opposite signs).

Integrating, gives

$$\cos^{-1}(x/A) = \omega t + D'.$$

But $x = A$ when $t = 0$, therefore $D' = 0$, so that

$$x(t) = A\cos(\omega t), \text{ where } A \text{ is the amplitude.} \tag{2.18}$$

3) If $a = h(v)$, then (2.19)

$$dv/dt = h(v)$$

therefore

$$dv/h(v) = dt,$$

and

$$\int dv/h(v) = t + B. \qquad (2.20)$$

Some of the techniques used to solve ordinary differential equations are discussed in Appendix A.

2.3 Velocity in Cartesian and polar coordinates

The transformation from Cartesian to Polar Coordinates is represented by the linear equations

$$x = r\cos\phi \text{ and } y = r\sin\phi, \qquad (2.21(\text{ a, b}))$$

or

$$x = f(r, \phi) \text{ and } y = g(r, \phi).$$

The differentials are

$$dx = (\partial f/\partial r)dr + (\partial f/\partial\phi)d\phi \text{ and } dy = (\partial g/\partial r)dr + (\partial g/\partial\phi)d\phi.$$

We are interested in the transformation of the components of the velocity vector under

$[x, y] \to [r, \phi]$. The velocity components involve the rates of change of dx and dy with respect to time:

$$dx/dt = (\partial f/\partial r)dr/dt + (\partial f/\partial\phi)d\phi/dt \text{ and } dy/dt = (\partial g/\partial r)dr/dt + (\partial g/\partial\phi)d\phi/dt$$

or

$$\dot{x} = (\partial f/\partial r)\dot{r} + (\partial f/\partial\phi)\dot{\phi} \text{ and } \dot{y} = (\partial g/\partial r)\dot{r} + (\partial g/\partial\phi)\dot{\phi}. \qquad (2.22)$$

But,

$$\partial f/\partial r = \cos\phi,\ \partial f/\partial\phi = -r\sin\phi,\ \partial g/\partial r = \sin\phi, \text{ and } \partial g/\partial\phi = r\cos\phi,$$

therefore, the velocity transformations are

$$\dot{x} = \cos\phi\,\dot{r} - \sin\phi(r\,\dot{\phi}) = v_x \qquad (2.23)$$

and

$$\dot{y} = \sin\phi\,\dot{r} + \cos\phi(r\,\dot{\phi}) = v_y. \qquad (2.24)$$

These equations can be written

$$\begin{pmatrix} v_x \\ v_y \end{pmatrix} = \begin{pmatrix} \cos\phi & -\sin\phi \\ \sin\phi & \cos\phi \end{pmatrix} \begin{pmatrix} dr/dt \\ r\,d\phi/dt \end{pmatrix} .$$

Changing $\phi \rightarrow -\phi$, gives the inverse equations

$$\begin{pmatrix} dr/dt \\ r\,d\phi/dt \end{pmatrix} = \begin{pmatrix} \cos\phi & \sin\phi \\ -\sin\phi & \cos\phi \end{pmatrix} \begin{pmatrix} v_x \\ v_y \end{pmatrix}$$

or

$$\begin{pmatrix} v_r \\ v_\phi \end{pmatrix} = \Re_c(\phi) \begin{pmatrix} v_x \\ v_y \end{pmatrix} . \qquad (2.25)$$

The velocity components in [r, ϕ] coordinates are therefore

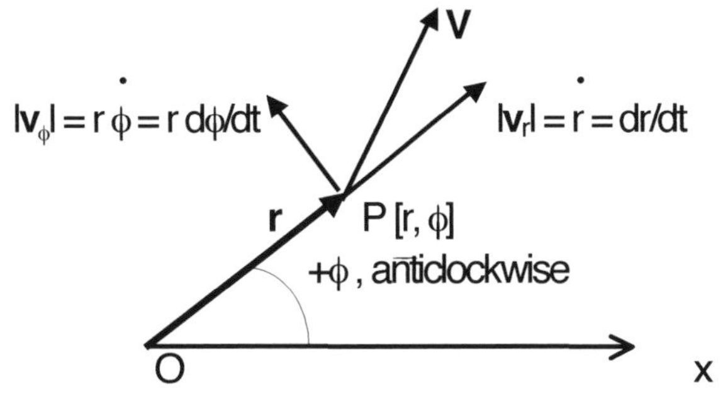

The quantity $d\phi/dt$ is called the *angular velocity* of P about the origin O.

2.4 Acceleration in Cartesian and polar coordinates

We have found that the velocity components transform from [x, y] to [r, ϕ] coordinates as follows

$$v_x = \cos\phi\,\dot{r} - \sin\phi(r\,\dot{\phi}) = \dot{x}$$

and

$$v_y = \sin\phi\,\dot{r} + \cos\phi(r\,\dot{\phi}) = \dot{y}.$$

The acceleration components are given by

$$a_x = dv_x/dt \text{ and } v_y = dv_y/dt$$

We therefore have

$$a_x = (d/dt)\{\cos\phi\,\dot{r} - \sin\phi(r\,\dot{\phi})\} \qquad (2.26)$$

$$= \cos\phi(\ddot{r} - r\,\dot{\phi}^2) - \sin\phi(2\dot{r}\,\dot{\phi} + r\,\ddot{\phi})$$

and

$$a_y = (d/dt)\{\sin\phi\,\dot{r} + \cos\phi(\dot{r}\,\dot{\phi})\} \qquad (2.27)$$

$$= \cos\phi(2\dot{r}\,\dot{\phi} + r\,\ddot{\phi}) + \sin\phi(\ddot{r} - r\,\dot{\phi}^2).$$

These equations can be written

$$\begin{pmatrix} a_r \\ a_\phi \end{pmatrix} = \begin{pmatrix} \cos\phi & \sin\phi \\ -\sin\phi & \cos\phi \end{pmatrix} \begin{pmatrix} a_x \\ a_y \end{pmatrix}. \qquad (2.28)$$

The acceleration components in [r, φ] coordinates are therefore

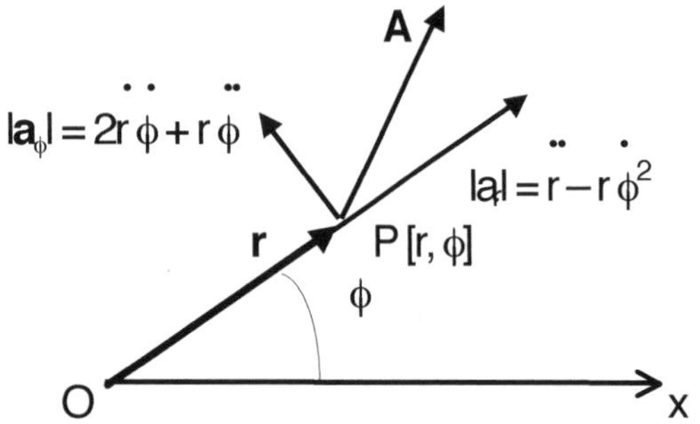

These expressions for the components of acceleration will be of key importance in discussions of Newton's Theory of Gravitation.

We note that, if r is constant, and the angular velocity ω is constant then

$$a_\phi = r\,\ddot{\phi} = r\,\dot{\omega} = 0, \qquad (2.29)$$

$$a_r = -r\,\dot{\phi}^2 = -r\,\omega^2 = -r\,(v_\phi/r)^2 = -v_\phi^{\,2}/r, \qquad (2.30)$$

and

$$v_\phi = r\dot{\phi} = r\,\omega. \quad (2.31)$$

These equations are true for *circular* motion.

PROBLEMS

2-1 A point moves with constant acceleration, a, along the x-axis. If it moves distances Δx_1 and Δx_2 in successive intervals of time Δt_1 and Δt_2, prove that the acceleration is

$$a = 2(v_2 - v_1)/T$$

where $v_1 = \Delta x_1/\Delta t_1$, $v_2 = \Delta x_2/\Delta t_2$, and $T = \Delta t_1 + \Delta t_2$.

2-2 A point moves along the x-axis with an instantaneous deceleration (negative acceleration):

$$a(t) \propto -v^{n+1}(t)$$

where v(t) is the instantaneous speed at time t, and n is a positive integer. If the initial speed of the point is u (at t = 0), show that

$$k_n t = \{(u^n - v^n)/(uv)^n\}/n,$$ where k_n is a constant of proportionality,

and that the distance traveled, x(t), by the point from its initial position is

$$k_n x(t) = \{(u^{n-1} - v^{n-1})/(uv)^{n-1}\}/(n-1).$$

2-3 A point moves along the x-axis with an instantaneous deceleration $kv^3(t)$, where v(t) is the speed and k is a constant. Show that

$$v(t) = u/(1 + k\,u\,x(t))$$

where x(t) is the distance travelled, and u is the initial speed of the point.

2-4 A point moves along the x-axis with an instantaneous acceleration

$$d^2x/dt^2 = -\omega^2/x^2$$

where ω is a constant. If the point starts from rest at $x = a$, show that the speed of the particle is

$$dx/dt = -\omega\{2(a - x)/(ax)\}^{1/2}.$$

Why is the negative square root chosen?

2-5 A point P moves with constant speed v along the x-axis of a Cartesian system, and a point Q moves with constant speed u along the y-axis. At time $t = 0$, P is at $x = 0$, and Q, moving towards the origin, is at $y = D$. Show that the minimum distance, d_{min}, between P and Q during their motion is

$$d_{min} = D\{1/(1 + (u/v)^2)\}^{1/2}.$$

Solve this problem in two ways:1) by direct minimization of a function, and 2) by a geometrical method that depends on the choice of a more suitable frame of reference (for example, the rest frame of P).

2-6 Two ships are sailing with constant velocities **u** and **v** on straight courses that are inclined at an angle θ. If, at a given instant, their distances from the point of intersection of their courses are a and b, find their minimum distance apart.

2-7 A point moves along the x-axis with an acceleration $a(t) = kt^2$, where t is the time the point has been in motion, and k is a constant. If the initial speed of the point is u, show that the distance travelled in time t is

$$x(t) = ut + (1/12)kt^4.$$

2-8 A point, moving along the x-axis, travels a distance x(t) given by the equation

$$x(t) = a\exp\{kt\} + b\exp\{-kt\}$$

where a, b, and k are constants. Prove that the acceleration of the point is proportional to the distance travelled.

2-9 A point moves in the plane with the equations of motion

$$\begin{pmatrix} d^2x/dt^2 \\ d^2y/dt^2 \end{pmatrix} = \begin{pmatrix} -2 & 1 \\ 1 & -2 \end{pmatrix} \begin{pmatrix} x \\ y \end{pmatrix}.$$

Let the following coordinate transformation be made

$$u = (x + y)/2 \text{ and } v = (x - y)/2.$$

Show that in the u-v frame, the equations of motion have a simple form, and that the time-dependence of the coordinates is given by

$$u = A\cos t + B\sin t,$$

and

$$v = C\cos\sqrt{3}\,t + D\sin\sqrt{3}\,t, \text{ where A, B, C, D are constants.}$$

This coordinate transformation has "diagonalized" the original matrix:

$$\begin{pmatrix} -2 & 1 \\ 1 & -2 \end{pmatrix} \rightarrow \begin{pmatrix} -1 & 0 \\ 0 & -3 \end{pmatrix}.$$

The matrix with zeros everywhere, except along the main diagonal, has the interesting property that it simply *scales* the vectors on which it acts — it does not rotate them. The scaling values are given by the diagonal elements, called the eigenvalues of the diagonal matrix. The scaled vectors are called eigenvectors. A small industry exists that is devoted to finding optimum ways of diagonalizing large matrices. Illustrate the motion of the system in the x-y frame and in the u-v frame.

3

CLASSICAL AND SPECIAL RELATIVITY

3.1 The Galilean transformation

Events belong to the physical world — they are not abstractions. We shall, nonetheless, introduce the idea of an *ideal* event that has neither extension nor duration. Ideal events may be represented as points in a space-time geometry. An event is described by a four-vector **E**[t, x, y, z] where t is the time, and x, y, z are the spatial coordinates, referred to arbitrarily chosen origins.

Let an event **E**[t, x], recorded by an observer O at the origin of an x-axis, be recorded as the event **E**′[t′, x′] by a second observer O′, moving at constant speed V along the x-axis. We suppose that their clocks are synchronized at $t = t' = 0$ when they coincide at a common origin, $x = x' = 0$.

At time t, we write the plausible equations

$$t' = t$$

and

$$x' = x - Vt,$$

where Vt is the distance traveled by O′ in a time t. These equations can be written

$$\mathbf{E}' = \mathbf{G}\mathbf{E} \qquad (3.1)$$

where

$$\mathbf{G} = \begin{pmatrix} 1 & 0 \\ -V & 1 \end{pmatrix}.$$

G is the operator of the Galilean transformation.

The inverse equations are

$$t = t'$$

and

$$x = x' + Vt'$$

or

$$\mathbf{E} = \mathbf{G}^{-1}\mathbf{E}' \tag{3.2}$$

where $\mathbf{G}^{-1}$ is the inverse Galilean operator (it undoes the effect of **G**).

If we multiply t and t′ by the constants k and k′, respectively, where k and k′ have dimensions of velocity then all terms have dimensions of length.

In space-space, we have the Pythagorean form $x^2 + y^2 = r^2$ (an invariant under rotations). We are therefore led to ask the question: is $(kt)^2 + x^2$ an invariant under **G** in space-time? Direct calculation gives

$$(kt)^2 + x^2 = (k't')^2 + x'^2 + 2Vx't' + V^2t'^2$$

$$= (k't')^2 + x'^2 \text{ only if } V = 0.$$

We see, therefore, that Galilean space-time does not leave the sum of squares invariant. We note, however, the key rôle played by *acceleration* in Galilean-Newtonian physics:

The velocities of the events according to O and O′ are obtained by differentiating

$x' = -Vt + x$ with respect to time, giving

$$v' = -V + v, \tag{3.3}$$

a result that agrees with everyday observations.

Differentiating v′ with respect to time gives

$$dv'/dt' = a' = dv/dt = a \tag{3.4}$$

where a and a′ are the accelerations in the two frames of reference. The classical acceleration is an *invariant* under the Galilean transformation. If the relationship $v' = v - V$ is used to describe the motion of a pulse of light, moving in empty space at $v = c \cong 3 \times 10^8$ m/s, it does not fit the facts. For example, if V is 0.5c, we expect to obtain $v' = 0.5c$ whereas, it is found that $v' = c$. In all cases studied, $v' = c$ for all values of V.

3.2 Einstein's space-time symmetry: the Lorentz transformation

It was Einstein, above all others, who advanced our understanding of the nature of space-time and relative motion. He made use of a symmetry argument to find the changes that must be made to the Galilean transformation if it is to account for the relative motion of rapidly moving objects and of beams of light. Einstein recognized an inconsistency in the Galilean-Newtonian equations, based as they are, on everyday experience. The discussion will be limited to non-accelerating, or so called inertial, frames

We have seen that the classical equations relating the events **E** and **E**′ are $\mathbf{E}' = \mathbf{GE}$, and the inverse $\mathbf{E} = \mathbf{G}^{-1}\mathbf{E}'$ where

$$\mathbf{G} = \begin{pmatrix} 1 & 0 \\ -V & 1 \end{pmatrix} \text{ and } \mathbf{G}^{-1} = \begin{pmatrix} 1 & 0 \\ V & 1 \end{pmatrix}.$$

These equations are connected by the substitution $V \leftrightarrow -V$; this is an algebraic statement of the Newtonian Principle of Relativity. Einstein incorporated this principle in his theory. He also retained the linearity of the classical equations in the absence of any evidence to the contrary. (Equispaced intervals of time and distance in one inertial frame remain equispaced in any other inertial frame). He *symmetrized* the space-time equations as follows:

$$\begin{pmatrix} t' \\ x' \end{pmatrix} = \begin{pmatrix} 1 & -V \\ -V & 1 \end{pmatrix} \begin{pmatrix} t \\ x \end{pmatrix}. \qquad (3.5)$$

Note, however, the inconsistency in the dimensions of the time-equation that has now been introduced:

$$t' = t - Vx.$$

The term Vx has dimensions of $[L]^2/[T]$, and not [T]. This can be corrected by introducing the *invariant* speed of light, c — a postulate in Einstein's theory that is consistent with the result of the Michelson-Morley experiment:

$$ct' = ct - Vx/c$$

so that all terms now have dimensions of length.

Einstein went further, and introduced a dimensionless quantity γ instead of the scaling factor of unity that appears in the Galilean equations of space-time. This factor must be consistent with all observations. The equations then become

$$ct' = \gamma ct - \beta\gamma x$$

$$x' = -\beta\gamma ct + \gamma x\text{, where } \beta = V/c.$$

These can be written

$$\mathbf{E}' = \mathbf{LE}, \tag{3.6}$$

where

$$\mathbf{L} = \begin{pmatrix} \gamma & -\beta\gamma \\ -\beta\gamma & \gamma \end{pmatrix},$$

and $\mathbf{E} = [ct, x]$.

L is the operator of the Lorentz transformation. The inverse equation is

$$\mathbf{E} = \mathbf{L}^{-1}\mathbf{E}' \tag{3.7}$$

where

$$\mathbf{L}^{-1} = \begin{pmatrix} \gamma & \beta\gamma \\ \beta\gamma & \gamma \end{pmatrix}.$$

This is the inverse Lorentz transformation, obtained from **L** by changing $\beta \rightarrow -\beta$ ($V \rightarrow -V$); it has the effect of undoing the transformation **L**. We can therefore write

$$\mathbf{L}\mathbf{L}^{-1} = \mathbf{I} \qquad (3.8)$$

Carrying out the matrix multiplications, and equating elements gives

$$\gamma^2 - \beta^2\gamma^2 = 1$$

therefore,

$$\gamma = 1/\sqrt{(1 - \beta^2)} \text{ (taking the positive root).} \qquad (3.9)$$

As $V \rightarrow 0$, $\beta \rightarrow 0$ and therefore $\gamma \rightarrow 1$; this represents the classical limit in which the Galilean transformation is, for all practical purposes, valid. In particular, time and space intervals have the same measured values in all Galilean frames of reference, and *acceleration* is the single fundamental invariant.

3.3 The invariant interval: contravariant and covariant vectors

Previously, it was shown that the space-time of Galileo and Newton is not Pythagorean under **G**. We now ask the question: is Einsteinian space-time Pythagorean under **L** ? Direct calculation leads to

$$\begin{aligned}(ct)^2 + x^2 &= \gamma^2(1 + \beta^2)(ct')^2 + 4\beta\gamma^2 x'ct' \\ &\quad + \gamma^2(1 + \beta^2)x'^2 \\ &\neq (ct')^2 + x'^2 \text{ if } \beta > 0.\end{aligned}$$

Note, however, that the *difference of squares is an invariant*:

$$(ct)^2 - x^2 = (ct')^2 - x'^2 \qquad (3.10)$$

because

$$\gamma^2(1 - \beta^2) = 1.$$

Space-time is said to be pseudo-Euclidean. The negative sign that characterizes Lorentz invariance can be included in the theory in a general way as follows.

We introduce two kinds of 4-vectors

$$x^{\mu} = [x^0, x^1, x^2, x^3], \text{ a } \textit{contravariant vector}, \tag{3.11}$$

and

$x_{\mu} = [x_0, x_1, x_2, x_3]$, a *covariant vector*, where

$$x_{\mu} = [x^0, -x^1, -x^2, -x^3]. \tag{3.12}$$

The scalar (or inner) product of the vectors is defined as

$$x^{\mu T}x_{\mu} = (x^0, x^1, x^2, x^3)[x^0, -x^1, -x^2, -x^3], \text{ to conform to matrix multiplication}$$

↑ row ↑ column

$$= (x^0)^2 - ((x^1)^2 + (x^2)^2 + (x^3)^2). \tag{3.13}$$

The superscript T is usually omitted in writing the invariant; it is implied in the form $x^{\mu}x_{\mu}$.

The event 4-vector is

$E^{\mu} = [ct, x, y, z]$ and the covariant form is

$E_{\mu} = [ct, -x, -y, -z]$

so that the invariant scalar product is

$$E^{\mu}E_{\mu} = (ct)^2 - (x^2 + y^2 + z^2). \tag{3.14}$$

A general Lorentz 4-vector x^{μ} transforms as follows:

$$x'^{\mu} = \mathbf{L}x^{\mu} \tag{3.15}$$

where

$$\mathbf{L} = \begin{pmatrix} \gamma & -\beta\gamma & 0 & 0 \\ -\beta\gamma & \gamma & 0 & 0 \\ 0 & 0 & 1 & 0 \\ 0 & 0 & 0 & 1 \end{pmatrix}$$

This is the operator of the Lorentz transformation if the motion of O′ is along the x-axis of O's frame of reference, and the initial times are synchronized ($t = t' = 0$ at $x = x' = 0$).

Two important consequences of the Lorentz transformation, discussed in **3.5**, are that intervals of time measured in two different inertial frames are not the same; they are related by the equation

$$\Delta t' = \gamma \Delta t \qquad (3.16)$$

where Δt is an interval measured on a clock at rest in O's frame, and distances are given by

$$\Delta l' = \Delta l / \gamma \qquad (3.17)$$

where Δl is a length measured on a ruler at rest in O's frame.

3.4 The group structure of Lorentz transformations

The square of the invariant interval s, between the origin [0, 0, 0, 0] and an arbitrary event $x^\mu = [x^0, x^1, x^2, x^3]$ is, in index notation

$$s^2 = x^\mu x_\mu = x'^\mu x'_\mu, \text{ (sum over } \mu = 0, 1, 2, 3). \qquad (3.18)$$

The lower indices can be raised using the metric tensor $\eta_{\mu\nu} = \text{diag}(1, -1, -1, -1)$, so that

$$s^2 = \eta_{\mu\nu} x^\mu x^\nu = \eta_{\mu\nu} x'^\mu x'^\nu, \text{ (sum over } \mu \text{ and } \nu). \qquad (3.19)$$

The vectors now have contravariant forms.

In matrix notation, the invariant is

$$s^2 = \mathbf{x}^T \eta \mathbf{x} = \mathbf{x}'^T \eta \mathbf{x}'. \qquad (3.20)$$

(The transpose must be written explicitly).

The primed and unprimed column matrices (contravariant vectors) are related by the Lorentz matrix operator, **L**

$$\mathbf{x}' = \mathbf{L}\mathbf{x}.$$

We therefore have

$$\mathbf{x}^T \eta \mathbf{x} = (\mathbf{L}\mathbf{x})^T \eta (\mathbf{L}\mathbf{x})$$

$$= \mathbf{x}^T \mathbf{L}^T \eta \mathbf{L} \mathbf{x}.$$

The $\mathbf{x}$'s are arbitrary, therefore

$$\mathbf{L}^{T}\eta\mathbf{L}=\eta. \tag{3.21}$$

This is the *defining* property of the Lorentz transformations.

The set of all Lorentz transformations is the set L of all 4 x 4 matrices that satisfies the defining property

$$L=\{\mathbf{L}: \mathbf{L}^{T}\eta\mathbf{L}=\eta;\ \mathbf{L} \text{ all 4 x 4 real matrices}; \eta=\text{diag}(1,-1,-1,-1\}.$$

(Note that each $\mathbf{L}$ has 16 (independent) real matrix elements, and therefore belongs to the 16-dimensional space, R^{16}).

Consider the result of two successive Lorentz transformations $\mathbf{L}_1$ and $\mathbf{L}_2$ that transform a 4-vector $\mathbf{x}$ as follows

$$\mathbf{x}\to\mathbf{x}'\to\mathbf{x}''$$

where

$$\mathbf{x}'=\mathbf{L}_1\mathbf{x},$$

and

$$\mathbf{x}''=\mathbf{L}_2\mathbf{x}'.$$

The resultant vector $\mathbf{x}''$ is given by

$$\begin{aligned}\mathbf{x}''&=\mathbf{L}_2(\mathbf{L}_1\mathbf{x})\\&=\mathbf{L}_2\mathbf{L}_1\mathbf{x}\\&=\mathbf{L}_c\mathbf{x}\end{aligned}$$

where

$$\mathbf{L}_c=\mathbf{L}_2\mathbf{L}_1\ (\mathbf{L}_1 \text{ followed by } \mathbf{L}_2). \tag{3.22}$$

If the combined operation $\mathbf{L}_c$ is always a Lorentz transformation then it must satisfy

$$\mathbf{L}_c^{T}\eta\mathbf{L}_c=\eta.$$

We must therefore have

$$(\mathbf{L}_2\mathbf{L}_1)^T\eta(\mathbf{L}_2\mathbf{L}_1) = \eta$$

or

$$\mathbf{L}_1{}^T(\mathbf{L}_2{}^T\eta\mathbf{L}_2)\mathbf{L}_1 = \eta$$

so that

$$\mathbf{L}_1{}^T\eta\mathbf{L}_1 = \eta, \quad (\mathbf{L}_1, \mathbf{L}_2 \in L)$$

therefore

$$\mathbf{L}_c = \mathbf{L}_2\mathbf{L}_1 \in L. \qquad (3.23)$$

Any number of successive Lorentz transformations may be carried out to give a resultant that is itself a Lorentz transformation.

If we take the determinant of the defining equation of $\mathbf{L}$,

$$\det(\mathbf{L}^T\eta\mathbf{L}) = \det\eta$$

we obtain

$$(\det\mathbf{L})^2 = 1 \ \ (\det\mathbf{L} = \det\mathbf{L}^T)$$

so that

$$\det\mathbf{L} = \pm 1. \qquad (3.24)$$

Since the determinant of $\mathbf{L}$ is not zero, an inverse transformation $\mathbf{L}^{-1}$ exists, and the equation $\mathbf{L}^{-1}\mathbf{L} = \mathbf{I}$, the identity, is always valid.

Consider the inverse of the defining equation

$$(\mathbf{L}^T\eta\mathbf{L})^{-1} = \eta^{-1},$$

or

$$\mathbf{L}^{-1}\eta^{-1}(\mathbf{L}^T)^{-1} = \eta^{-1}.$$

Using $\eta = \eta^{-1}$, and rearranging, gives

$$\mathbf{L}^{-1}\eta(\mathbf{L}^{-1})^{T} = \eta . \tag{3.25}$$

This result shows that the inverse $\mathbf{L}^{-1}$ is always a member of the set L.

The Lorentz transformations **L** are matrices, and therefore they obey the associative property under matrix multiplication.

We therefore see that

1. If $\mathbf{L}_1$ and $\mathbf{L}_2 \in L$, then $\mathbf{L}_2\,\mathbf{L}_1 \in L$

2. If $\mathbf{L} \in L$, then $\mathbf{L}^{-1} \in L$

3. The identity $\mathbf{I} = \text{diag}(1, 1, 1, 1) \in L$, and

4. The matrix operators **L** obey associativity.

The set of all Lorentz transformations therefore forms a *group*.

3.5 The rotation group

Spatial rotations in two and three dimensions are Lorentz transformations in which the time-component remains unchanged. In Chapter **1**, the geometrical properties of the rotation operators are discussed. In this section, we shall consider the algebraic structure of the operators.

Let $\Re$ be a real 3×3 matrix that is part of a Lorentz transformation with a constant

time - component,

$$\mathbf{L} = \begin{pmatrix} 1 & 0 & 0 & 0 \\ 0 & & & \\ 0 & & \Re & \\ 0 & & & \end{pmatrix} . \tag{3.26}$$

In this case, the defining property of the Lorentz transformations leads to

$$\begin{pmatrix} 1 & 0 & 0 & 0 \\ 0 & & & \\ 0 & & \Re^{T} & \\ 0 & & & \end{pmatrix} \begin{pmatrix} 1 & 0 & 0 & 0 \\ 0 & -1 & 0 & 0 \\ 0 & 0 & -1 & 0 \\ 0 & 0 & 0 & -1 \end{pmatrix} \begin{pmatrix} 1 & 0 & 0 & 0 \\ 0 & & & \\ 0 & & \Re & \\ 0 & & & \end{pmatrix} = \begin{pmatrix} 1 & 0 & 0 & 0 \\ 0 & -1 & 0 & 0 \\ 0 & 0 & -1 & 0 \\ 0 & 0 & 0 & -1 \end{pmatrix} \tag{3.27}$$

so that

$$\mathfrak{R}^T\mathfrak{R} = \mathbf{I}\text{, the identity matrix, diag(1,1,1).}$$

This is the defining property of a three-dimensional orthogonal matrix. (The related two - dimensional case is treated in Chapter **1**).

If $\mathbf{x} = [x_1, x_2, x_3]$ is a three-vector that is transformed under $\mathfrak{R}$ to give $\mathbf{x}'$ then

$$\mathbf{x}'^T\mathbf{x}' = \mathbf{x}^T\mathfrak{R}^T\mathfrak{R}\mathbf{x} = \mathbf{x}^T\mathbf{x} = x_1^2 + x_2^2 + x_3^2 = \text{invariant under } \mathfrak{R}. \tag{3.28}$$

The action of $\mathfrak{R}$ on any three-vector preserves length. The set of all 3×3 orthogonal matrices is denoted by **O**(3),

$$\mathbf{O}(3) = \{\mathfrak{R} : \mathfrak{R}^T\mathfrak{R} = \mathbf{I},\ r_{ij} \in \text{Reals}\}.$$

The elements of this set satisfy the four group axioms.

3.6 The relativity of simultaneity: time dilation and length contraction

In order to record the time and place of a sequence of events in a particular inertial reference frame, it is necessary to introduce an infinite set of adjacent "observers", located throughout the entire space. Each observer, at a known, fixed position in the reference frame, carries a clock to record the time and the characteristic property of every event in his immediate neighborhood. The observers are not concerned with non-local events. The clocks carried by the observers are synchronized — they all read the same time throughout the reference frame. The process of synchronization is discussed later. It is the job of the chief observer to collect the information concerning the time, place, and characteristic feature of the events recorded by all observers, and to construct the world line (a path in space-time), associated with a particular characteristic feature (the type of particle, for example).

Consider two sources of light, 1 and 2, and a point M midway between them. Let E_1 denote the event "flash of light leaves 1", and E_2 denote the event "flash of light leaves 2". The events E_1 and E_2 are *simultaneous* if the flashes of light from 1 and 2 reach M at the same time. The fact that the speed of light in free space is independent of the speed of the source means that *simultaneity is relative.*

The clocks of all the observers in a reference frame are synchronized by correcting them for the speed of light as follows:

Consider a set of clocks located at x_0, x_1, x_2, x_3, ... along the x-axis of a reference frame. Let x_0 be the chief's clock, and let a flash of light be sent from the clock at x_0 when it is reading t_0 (12 noon, say). At the instant that the light signal reaches the clock at x_1, it is set to read $t_0 + (x_1/c)$, at the instant that the light signal reaches the clock at x_2, it is set to read $t_0 + (x_2/c)$, and so on for every clock along the x-axis. All clocks in the reference frame then "read the same time" — they are synchronized. From the viewpoint of all other inertial observers, in their own reference frames, the set of clocks, sychronized using the above procedure, appears to be unsychronized. It is the lack of symmetry in the sychronization of clocks in different reference frames that leads to two non-intuitive results namely, length contraction and time dilation.

Length contraction: an application of the Lorentz transformation.

Consider a rigid rod at rest on the x-axis of an inertial reference frame S´. Because it is at rest, it does not matter when its end-points x_1' and x_2' are measured to give the rest-, or proper-length of the rod, $L_0' = x_2' - x_1'$. Consider the same rod observed in an inertial reference frame S that is moving with constant velocity $-\mathbf{V}$ with its x-axis parallel to the x´-axis. We wish to determine the length of the moving rod; we require the length $L = x_2 - x_1$ according to the observers in S. This means that the observers in S must measure x_1 and x_2 *at the same time* in their reference frame. The events in the two reference frames S, and S´ are related by the spatial part of the Lorentz transformation:

$$x' = -\beta\gamma ct + \gamma x$$

and therefore

$$x_2' - x_1' = -\beta\gamma c(t_2 - t_1) + \gamma(x_2 - x_1).$$

where

$$\beta = V/c \text{ and } \gamma = 1/\sqrt{(1-\beta^2)}.$$

Since we require the length $(x_2 - x_1)$ in S to be measured at the same time in S, we must have $t_2 - t_1 = 0$, and therefore

$$L_0' = x_2' - x_1' = \gamma(x_2 - x_1),$$

or

$$L_0'(\text{at rest}) = \gamma L\ (\text{moving}). \qquad (3.29)$$

The length of a moving rod L is less than the length of the same rod measured at rest, L_0, because $\gamma > 1$.

Time dilation

Consider a clock at rest at the origin of an inertial frame S´, and a set of synchronized clocks at x_0, x_1, x_2, ... on the x-axis of another inertial frame S. Let S´ move at constant speed V relative to S, along the common x -, x´- axis. Let the clocks at x_o, and x_o' be sychronized to read t_0 , and t_0' at the instant that they coincide in space. A *proper time interval* is defined to be the time between two events measured in an inertial frame in which the two events occur at the same place. The time part of the Lorentz transformation can be used to relate an interval of time measured on the single clock in the S´ frame, and the same interval of time measured on the set of synchronized clocks at rest in the S frame. We have

$$ct = \gamma ct' + \beta\gamma x'$$

or

$$c(t_2 - t_1) = \gamma c(t_2' - t_1') + \beta\gamma(x_2' - x_1').$$

There is no separation between a single clock and itself, therefore $x_2' - x_1' = 0$, so that

$$c(t_2 - t_1)(\text{moving}) = \gamma c(t_2' - t_1')(\text{at rest}) \quad (\gamma > 1). \qquad (3.30)$$

A moving clock runs more slowly than a clock at rest.

In Chapter **1**, it was shown that the general 2×2 matrix operator transforms rectangular coordinates into oblique coordinates. The Lorentz transformation is a special case of the 2×2 matrices, and therefore its effect is to transform rectangular space-time coordinates into oblique space-time coordinates:

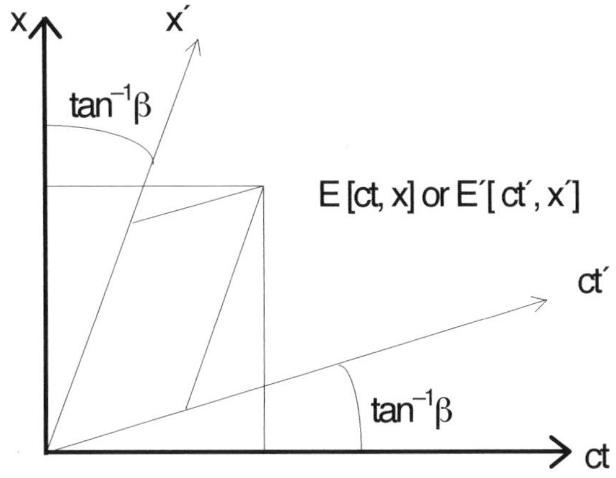

The geometrical form of the Lorentz transformation

The symmetry of space-time means that the transformed axes rotate through equal angles, $\tan^{-1}\beta$. The relativity of simultaneity is clearly exhibited on this diagram: two events that occur at the same time in the ct, x -frame necessarily occur at different times in the oblique ct′, x′-frame.

3.7 The 4-velocity

A differential time interval, dt, cannot be used in a Lorentz-invariant way in kinematics. We must use the proper time differential interval, $d\tau$, defined by

$$(cdt)^2 - dx^2 = (cdt')^2 - dx'^2 \equiv (cd\tau)^2. \qquad (3.31)$$

The Newtonian 3-velocity is

$$\mathbf{v}_N = [dx/dt, dy/dt, dz/dt],$$

and this must be replaced by the 4-velocity

$$V^\mu = [d(ct)/d\tau, dx/d\tau, dy/d\tau, dz/d\tau]$$

$= [d(ct)/dt, dx/dt, dy/dt, dz/dt](dt/d\tau)$

$= [\gamma c, \gamma \mathbf{v}_N]$. (3.32)

The scalar product is then

$V^\mu V_\mu = (\gamma c)^2 - (\gamma v_N)^2$ (the transpose is understood)

$= (\gamma c)^2(1 - (v_N/c)^2)$

$= c^2$. (3.33)

The magnitude of the 4-velocity is therefore $|V^\mu| = c$, the invariant speed of light.

PROBLEMS

3-1 Two points A and B move in the plane with constant velocities $|\mathbf{v}_A| = \sqrt{2}$ m.s^{-1} and $|\mathbf{v}_B| = 2\sqrt{2}$ m.s^{-1}. They move from their initial (t = 0) positions, A(0)[1, 1] and B(0)[6, 2] as shown:

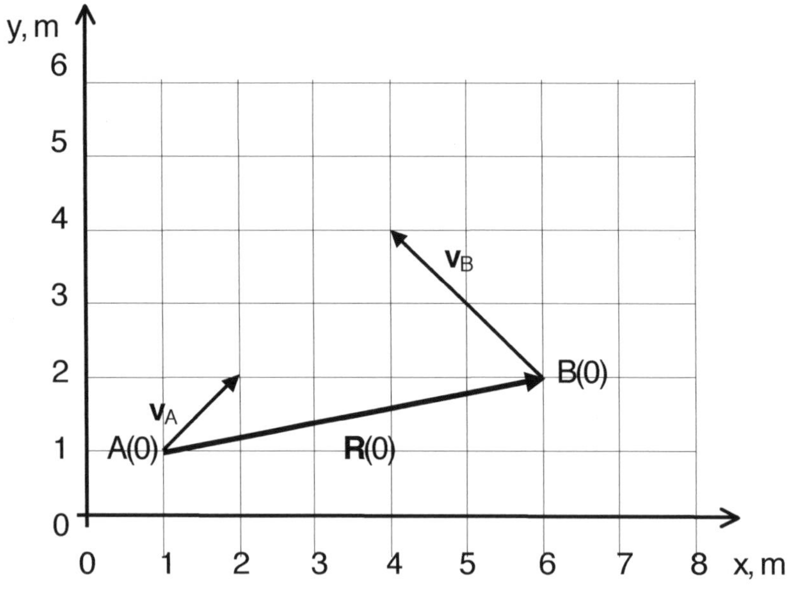

Show that the closest distance between the points is $|\mathbf{R}|_{min}$ = 2.529882..meters, and that it occurs 1.40...seconds after they leave their initial positions. (Remember that all inertial frames are equivalent, therefore choose the most appropriate for dealing with this problem).

3-2 Show that the set of all standard (motion along the common x-axis) Galilean transformations forms a group.

3-3 A flash of light is sent out from a point x_1 on the x-axis of an inertial frame S, and it is received at a point $x_2 = x_1 + l$. Consider another inertial frame, S´, moving with constant speed $V = \beta c$ along the x-axis; show that, in S´:

i) the separation between the point of emission and the point of reception of the light is $$l' = l\{(1 - \beta)/(1 + \beta)\}^{1/2}$$

ii) the time interval between the emission and reception of the light is

$$\Delta t' = (l/c)\{(1 - \beta)/(1 + \beta)\}^{1/2}$$

3-4 The distance between two photons of light that travel along the x-axis of an inertial frame, S, is always l. Show that, in a second inertial frame, S´, moving at constant speed $V = \beta c$ along the x-axis, the separation between the two phot ons is

$$\Delta x' = l\{(1 + \beta)/(1 - \beta)\}^{1/2}.$$

3-5 An event [ct, x] in an inertial frame, S, is transformed under a standard Lorentz transformation to [ct´, x´] in a standard primed frame, S´, that has a constant speed V along the x-axis, show that the velocity components of the point x, x´ are related by the equation

$$v_x = (v_x' + V)/(1 + (v_x' V/c^2)).$$

3-6 An object called a K^0-meson decays when at rest into two objects called π-mesons ($\pi^\pm$), each with a speed of 0.8c. If the K^0-meson has a measured speed of 0.9c when it decays, show that the greatest speed of one of the π-mesons is (85/86)c and that its least speed is (5/14).

4

NEWTONIAN DYNAMICS

Although our discussion of the geometry of motion has led to major advances in our understanding of measurements of space and time in different inertial systems, we have yet to come to the crux of the matter, namely — a discussion of the effects of *forces* on the motion of two or more interacting particles. This key branch of Physics is called Dynamics. It was founded by Galileo and Newton and perfected by their followers, most notably Lagrange and Hamilton. We shall see that the Newtonian concepts of momentum and kinetic energy require fundamental revisions in the light of the Einstein's Special Theory of Relativity. The revised concepts come about as a result of Einstein's recognition of the crucial rôle of the Principle of Relativity in unifying the dynamics of *all* mechanical and optical phenomena. In spite of the conceptual difficulties inherent in the classical concepts, (difficulties that will be discussed later), the subject of Newtonian dynamics represents one of the great triumphs of Natural Philosophy. The successes of the classical theory range from accurate descriptions of the dynamics of everyday objects to a detailed understanding of the motions of galaxies.

4.1 The law of inertia

Galileo (1544-1642) was the first to develop a quantitative approach to the study of motion. He addressed the question — what property of motion is related to force? Is it the position of the moving object? Is it the velocity of the moving object? Is it the rate of change of its velocity? ...The answer to the question can be obtained only from observations; this is a basic feature of Physics that sets it apart from Philosophy proper. Galileo observed that *force* influences the *changes in velocity* (accelerations) of an object and that, in the absence of external forces (e.g: friction), no force is needed to keep an object in motion that is traveling in a straight line with constant speed. This observationally based law is called the *Law of Inertia*. It is, perhaps, difficult for us to appreciate the impact of Galileo's new ideas concerning motion. The fact that an object resting on a horizontal surface remains at rest unless something we call force is applied to change its state of rest was, of course, well-known before Galileo's time. However, the fact that the object continues to move after the force ceases to be applied caused considerable conceptual difficulties for the early Philosophers (see Feynman *The Character of Physical Law*). The observation that, in practice, an object comes to rest due to frictional forces and air resistance was recognized by Galileo to be a side effect, and not germane to the fundamental question of motion. Aristotle, for example, believed that the true or natural state of motion is one of rest. It is instructive to consider Aristotle's conjecture from the viewpoint of the Principle of Relativity —- is a natural state of rest consistent with this general Principle? According to the general Principle of Relativity, the laws of motion have the same form in all frames of reference that move with constant speed in straight lines with respect to each other. An observer in a reference frame moving with constant speed in a straight line with respect to the reference frame in which the object is at rest would conclude that the natural state or motion of the object is one of constant speed in a straight line, and not one of rest. All inertial observers, in an infinite number of frames of reference, would come to the same conclusion. We see, therefore, that Aristotle's conjecture is not consistent with this fundamental Principle.

4.2 Newton's laws of motion

During his early twenties, Newton postulated three Laws of Motion that form the basis of Classical Dynamics. He used them to solve a wide variety of problems including the dynamics of the planets. The Laws of Motion, first published in the *Principia* in 1687, play a fundamental rôle in Newton's Theory of Gravitation (Chapter **7**); they are:

1. In the *absence* of an applied force, an object will remain at rest or in its present state of constant speed in a straight line (Galileo's Law of Inertia)

2. In the *presence* of an applied force, an object will be accelerated in the direction of the applied force and the product of its mass multiplied by its acceleration is equal to the force.

and,

3. If a body A exerts a force of magnitude $|\mathbf{F}_{AB}|$ on a body B, then B exerts a force of equal magnitude $|\mathbf{F}_{BA}|$ on A.. The forces act in opposite directions so that

$$\mathbf{F}_{AB} = -\mathbf{F}_{BA} .$$

In law number 2, the acceleration lasts only while the applied force lasts. The applied force need not, however, be constant in time — the law is true at all times during the motion. Law number 3 applies to "contact" interactions. If the bodies are separated, and the interaction takes a finite time to propagate between the bodies, the law must be modified to include the properties of the "field " between the bodies. This important point is discussed in Chapter **7**.

4.3 Systems of many interacting particles: conservation of linear and angular momentum

Studies of the dynamics of two or more interacting particles form the basis of a key part of Physics. We shall deduce two fundamental principles from the Laws of Motion; they are:

1) The *Conservation of Linear Momentum* which states that, if there is a direction in which the sum of the components of the external forces acting on a system is zero, then the linear momentum of the system in that direction is constant, and

2) The *Conservation of Angular Momentum* which states that, if the sum of the moments of the external forces about any fixed axis (or origin) is zero, then the angular momentum about that axis (or origin) is constant.

The new terms that appear in these statements will be defined later.

The first of these principles will be deduced by considering the dynamics of two interacting particles of masses m_1 and m_2 with instantaneous coordinates $[x_1, y_1]$ and $[x_2, y_2]$, respectively. In Chapter **12**, these principles will be deduced by considering the invariance of the Laws of Motion under translations and rotations of the coordinate systems.

Let the external forces acting on the particles be $\mathbf{F}_1$ and $\mathbf{F}_2$, and let the mutual interactions be $\mathbf{F}_{21}'$ and $\mathbf{F}_{12}'$. The system is as shown

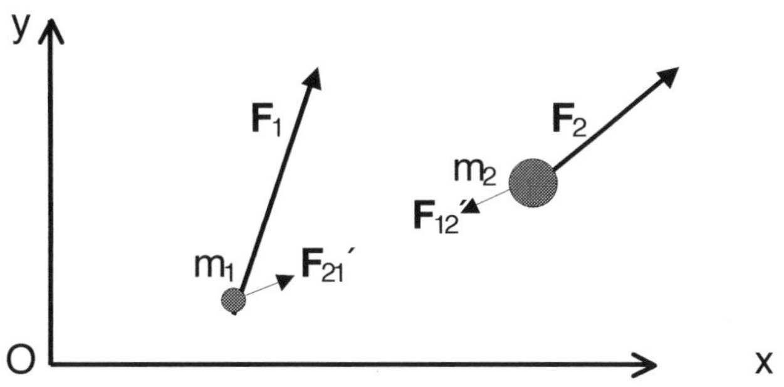

Resolving the forces into their x- and y-components gives

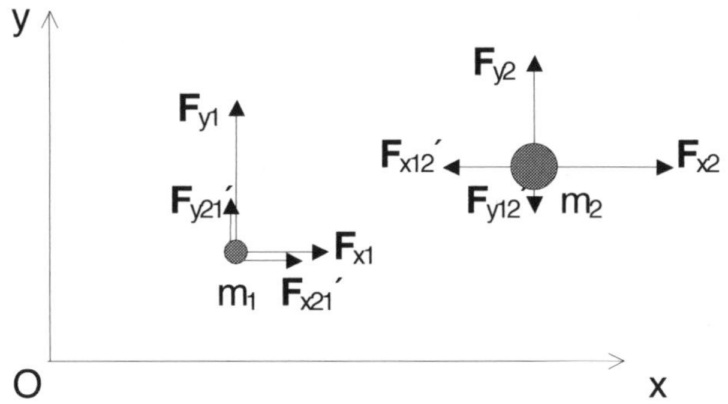

a) *The equations of motion*

The equations of motion for each particle are

1) Resolving in the x-direction

$$F_{x1} + F_{x21}' = m_1 (d^2x_1/dt^2) \tag{4.1}$$

and

$$F_{x2} - F_{x12}' = m_2(d^2x_2/dt^2). \tag{4.2}$$

Adding these equations gives

$$F_{x1} + F_{x2} + (F_{x21}' - F_{x12}') = m_1(d^2x_1/dt^2) + m_2(d^2x_2/dt^2). \quad (4.3)$$

2) Resolving in the y-direction gives a similar equation, namely

$$F_{y1} + F_{y2} + (F_{y21}' - F_{y12}') = m_1(d^2y_1/dt^2) + m_2(d^2y_2/dt^2). \quad (4.4)$$

b) *The rôle of Newton's 3rd Law*

For instantaneous mutual interactions, Newton's 3rd Law gives $|\mathbf{F}_{21}'| = |\mathbf{F}_{12}'|$

so that the x- and y-components of the internal forces are themselves equal and opposite, therefore the total equations of motion are

$$F_{x1} + F_{x2} = m_1(d^2x_1/dt^2) + m_2(d^2x_2/dt^2), \quad (4.5)$$

and

$$F_{y1} + F_{y2} = m_1(d^2y_1/dt^2) + m_2(d^2y_2/dt^2). \quad (4.6)$$

c) *The conservation of linear momentum*

If the sum of the external forces acting on the masses in the x-direction is zero, then

$$F_{x1} + F_{x2} = 0, \quad (4.7)$$

in which case,

$$0 = m_1(d^2x_1/dt^2) + m_2(d^2x_2/dt^2)$$

or

$$0 = (d/dt)(m_1v_{x1}) + (d/dt)(m_2v_{x2}),$$

which, on integration gives

$$\text{constant} = m_1v_{x1} + m_2v_{x2}\,. \quad (4.8)$$

The product (mass × velocity) is the linear momentum. We therefore see that if there is no resultant external force in the x-direction, the linear momentum of the two particles in the x-direction is conserved. The above argument can be generalized so that we can state: the linear momentum of the two particles is constant in any direction in which there is no resultant external force.

4.3.1 Interaction of n-particles

The analysis given in **4.3** can be carried out for an arbitrary number of particles, n, with masses m_1, m_2, ...m_n and with instantaneous coordinates $[x_1, y_1]$, $[x_2, y_2]$ ..$[x_n, y_n]$. The mutual interactions cancel in pairs so that the equations of motion of the n-particles are, in the x-direction

$$F_{x1} + F_{x2} + \ldots F_{xn} = m_1\ddot{x}_1 + m_2\ddot{x}_2 + \ldots m_n\ddot{x}_n = \text{sum of the x-components of the external forces acting on the masses,} \tag{4.9}$$

and, in the y-direction

$$F_{y1} + F_{y2} + \ldots F_{yn} = m_1\ddot{y}_1 + m_1\ddot{y}_2 + \ldots m_n\ddot{y}_n = \text{sum of the y-components of the external forces acting on the masses.} \tag{4.10}$$

In this case, we see that if the sum of the components of the external forces acting on the system in a particular direction is zero, then the linear momentum of the system in that direction is constant. If, for example, the direction is the x-axis then

$$m_1v_{x1} + m_2v_{x2} + \ldots m_nv_{xn} = \text{constant.} \tag{4.11}$$

4.3.2 Rotation of two interacting particles about a fixed point

We begin the discussion of the second fundamental conservation law by considering the motion of two interacting particles that move under the influence of external forces $\mathbf{F}_1$ and $\mathbf{F}_2$, and mutual interactions (internal forces) $\mathbf{F}_{21}'$ and $\mathbf{F}_{12}'$. We are interested in the motion of the two masses about a fixed point O that is chosen to be the origin of Cartesian coordinates. The perpendiculars drawn from the point O to the lines of action of the forces are R_1, R_2, and R'. The system is illustrated in the following figure.

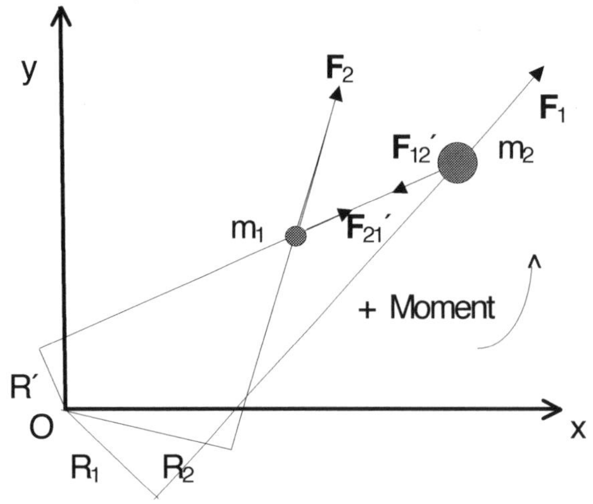

a) *The moment of forces about a fixed origin*

The total moment $\Gamma_{1,2}$ of the forces about the origin O is defined as

$$\Gamma_{1,2} = \underbrace{R_1F_1 + R_2F_2}_{\text{moment of external forces}} + \underbrace{(R'F_{12}' - R'F_{21}')}_{\text{moment of internal forces}} \tag{4.12}$$

A positive moment acts in a counter-clockwise sense.

Newton's 3rd Law gives

$$|\mathbf{F}_{21}'| = |\mathbf{F}_{12}'|,$$

therefore, the moment of the internal forces about O is zero. (Their lines of action are the same).

The total effective moment about O is therefore due to the external forces, alone. Writing the moment in terms of the x- and y-components of $\mathbf{F}_1$ and $\mathbf{F}_2$, we obtain

$$\Gamma_{1,2} = x_1F_{y1} + x_2F_{y2} - y_1F_{x1} - y_2F_{x2} \tag{4.13}$$

b) *The conservation of angular momentum*

If the moment of the external forces about the origin O is zero then, by integration, we have

$$\text{constant} = x_1p_{y1} + x_2p_{y2} - y_1p_{x1} - y_2p_{x2}.$$

where p_{x1} is the x-component of the momentum of mass 1, etc..

Rearranging, gives

$$\text{constant} = (x_1p_{y1} - y_1p_{x1}) + (x_2p_{y2} - y_2p_{x2}). \qquad (4.14)$$

The right-hand side of this equation is called the *angular momentum of the two particles about the fixed origin, O.*

Alternatively, we can discuss the conservation of angular momentum using vector analysis. Consider a non-relativistic particle of mass m and momentum **p**, moving in the plane under the influence of an external force **F** about a fixed origin, O:

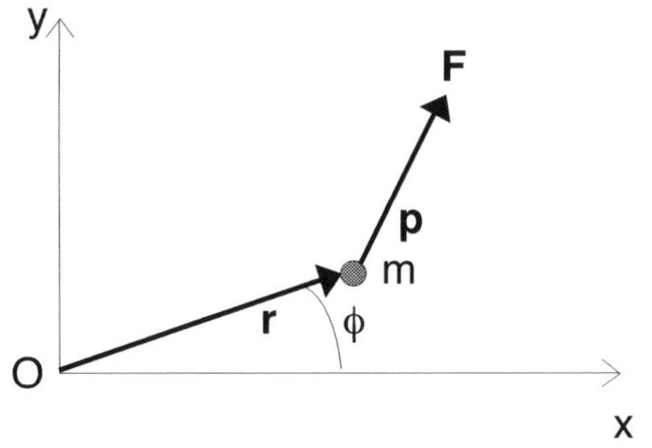

The angular momentum, **L**, of m about O can be written in vector form

$$\mathbf{L} = \mathbf{r} \times \mathbf{p}. \qquad (4.15)$$

The *torque*, Γ, associated with the external force **F** acting about O is

$$\Gamma = \mathbf{r} \times \mathbf{F}. \qquad (4.16)$$

The rate of change of the angular momentum with time is

$$d\mathbf{L}/dt = \mathbf{r} \times (d\mathbf{p}/dt) + \mathbf{p} \times (d\mathbf{r}/dt) \qquad (4.17)$$

$$= \mathbf{r} \times m(d\mathbf{v}/dt) + m\mathbf{v} \times \mathbf{v}$$

$$= \mathbf{r} \times \mathbf{F} \text{ (because } \mathbf{v} \times \mathbf{v} = 0)$$

$$= \Gamma.$$

If there is no external torque, $\Gamma = 0$. We have, therefore

$$\Gamma = d\mathbf{L}/dt = 0, \tag{4.18}$$

so that **L** is a constant of the motion.

4.3.3 Rotation of n-interacting particles about a fixed point

The analysis given in **4.3.2** can be extended to a system of n-interacting particles. The moments of the mutual interactions about the origin O cancel in pairs (Newton's 3rd Law) so that we are left with the moment of the external forces about O. The equation for the total moment is therefore

$$\Gamma_{1,2,\ldots n} = \sum_{[i=1,n]} (x_i d(m_i v_{yi})/dt - y_i d(m_i v_{xi})/dt). \tag{4.19}$$

If the moment of the external forces about the fixed origin is zero then the total angular momentum of the system about O is a constant. This result follows directly by integrating the expression for $\Gamma_{1,2,\ldots n} = 0$.

If the origin moves with constant velocity, the angular momentum of the system, relative to the new coordinate system, is constant if the external torque is zero.

4.4 Work and energy in Newtonian dynamics

4.4.1 The principle of work: kinetic energy and the work done by forces

Consider a mass m moving along a path in the [x, y]-plane under the influence of a resultant force **F** that is not necessarily constant. Let the components of the force be $\mathbf{F}_x$ and $\mathbf{F}_y$ when the mass is at the point P[x, y]. We wish to study the motion of m in moving from a point A[x_A, y_A] where the force is $\mathbf{F}_A$ to a point B[x_B, y_B] where the force is $\mathbf{F}_B$. The equations of motion are

$$m(d^2x/dt^2) = F_x \tag{4.20}$$

and

$$m(d^2y/dt^2) = F_y \tag{4.21}$$

Multiplying these equations by dx/dt and dy/dt, respectively, and adding, we obtain

$$m(dx/dt)(d^2x/dt^2) + m(dy/dt)(d^2y/dt^2) = F_x(dx/dt) + F_y(dy/dt).$$

This equation now can be integrated with respect to t, so that

$$m((dx/dt)^2 + (dy/dt)^2)/2 = \int (F_x dx + F_y dy) .$$

or

$$mv^2/2 = \int (F_x dx + F_y dy), \qquad (4.22)$$

where $v = ((dx/dt)^2 + (dy/dt)^2)^{1/2}$ is the speed of the particle at the point [x, y]. The term $mv^2/2$ is called the *classical kinetic energy* of the mass m. It is important to note that the kinetic energy is a *scalar*.

If the resultant forces acting on m are $\mathbf{F}_A$ at $A[x_A, y_A]$ at time t_A, and $\mathbf{F}_B$ at $B[x_B, y_B]$ at time t_B, then we have

$$mv_B^2/2 - mv_A^2/2 = \int_{[xA, xB]} F_x dx + \int_{[yA, yB]} F_y dy . \qquad (4.23)$$

The terms on the right-hand side of this equation represent the *work done* by the resultant forces acting on the particle in moving it from A to B. The equation is the mathematical form of the general *Principle of Work*: the change in the kinetic energy of a system in any interval of time is equal to the work done by the resultant forces acting on the system during that interval.

4.5 Potential energy

4.5.1 General features

Newtonian dynamics involves vector quantities — force, momentum, angular momentum, etc.. There is, however, another form of dynamics that involves scalar quantities; a form that originated in the works of Huygens and Leibniz, in the 17th century. The scalar form relies upon the concept of energy, in its broadest sense. We have met the concept of kinetic energy in the previous section. We now meet a more abstract quantity called *potential energy*.

The work done, W, by a force, **F**, in moving a mass m from a position s_A to a position s_B along a path s is, from section **4.3**,

$W = \int_{[sA,\, sB]} \mathbf{F} \cdot d\mathbf{s}$ = the change in the kinetic energy during the motion,

$= \int_{[sA,\, sB]} F\, ds \cos\alpha$, where α is the angle between **F** and d**s**. (4.24)

If the force is constant, we can write

$$W = F(s_B - s_A),$$

where $s_B - s_A$ is the arc length.

If the motion is along the x-axis, and $F = F_x$ is constant then

$W = F_x(x_B - x_A)$, the force multiplied by the distance moved. (4.25)

This equation can be rearranged, as follows

$$mv_{xB}^2/2 - F_x x_B = mv_{xA}^2/2 - F_x x_A\,. \qquad (4.26)$$

This is a surprising result; the kinetic energy of the mass *is not conserved* during the motion whereas the quantity $(mv_x^2/2 - F_x x)$ *is conserved* during the motion. This means that the change in the kinetic energy is exactly balanced by the change in the quantity $F_x x$.

Since the quantity $mv^2/2$ has dimensions of energy, the quantity $F_x x$ must have dimensions of energy if the equation is to be dimensionally correct. The quantity $-F_x x$ is called the potential energy of the mass m, when at the position x, due to the influence of the force F_x. We shall denote the potential energy by V. The negative sign that appears in the definition of the potential energy will be discussed later when explicit reference is made to the nature of the force (for example, gravitational or electromagnetic).

The energy equation can therefore be written

$$T_B + V_B = T_A + V_A\,. \qquad (4.27)$$

This is found to be a general result that holds in all cases in which a potential energy function can be found that depends only on the position of the object (or objects).

4.5.2 Conservative forces

Let $\mathbf{F}_x$ and $\mathbf{F}_y$ be the Cartesian components of the forces acting on a moving particle with coordinates [x, y]. The work done $W_{1\to 2}$ by the forces while the particle moves from the position $P_1\,[x_1, y_1]$ to another position $P_2[\,x_2, y_2]$ is

$$W_{1\to 2} = \int_{[x1,\,x2]} F_x dx + \int_{[y1,\,y2]} F_y dy \tag{4.28}$$

$$= \int_{[P1,\,P2]} (F_x dx + F_y dy)\,.$$

If the quantity $F_x dx + F_y dy$ is a perfect differential then a function $U = f(x, y)$ exists such that

$$F_x = \partial U/\partial x \text{ and } F_y = \partial U/\partial y\,. \tag{4.29}$$

Now, the *total* differential of the function U is

$$dU = (\partial U/\partial x)dx + (\partial U/\partial y)dy \tag{4.30}$$

$$= F_x dx + F_y dy.$$

In this case, we can write

$$\int dU = \int(F_x dx + F_y dy) = U = f(x, y).$$

The definite integral evaluated between $P_1\,[x_1, y_1]$ and $P_2[x_2, y_2]$ is

$$\int_{[P1,\,P2]} (F_x dx + F_y dy) = f(x_2, y_2) - f(x_1, y_1) = U_2 - U_1\,. \tag{4.31}$$

We see that in evaluating the work done by the forces during the motion, no mention is made of the actual path taken by the particle. If the forces are such that the function U(x, y) exists, then they are said to be *conservative*. The function U(x, y) is called the *force function*.

The above method of analysis can be applied to a system of many particles, n. The total work done by the resultant forces acting on the system in moving the particles from their initial configuration, i, to their final configuration, f, is

$$W_{i\to f} = \sum_{[k=1,\,n]} \int_{[Pk1,\,Pk2]} (F_{kx} dx_k + F_{ky} dy_k), \tag{4.32}$$

$= U_f - U_i,$

a scalar quantity that is independent of the paths taken by the individual particles. $P_{k1}[x_{k1}, y_{k1}]$ and $P_{k2}[x_{k2}, y_{k2}]$ are the initial and final coordinates of the kth-particle.

The potential energy, V, of the system moving under the influence of conservative forces is defined in terms of the function U: $V \equiv -U$.

Examples of interactions that take place via conservative forces are:

1) gravitational interactions

2) electromagnetic interactions

and

3) interactions between particles of a system that, for every pair of particles, act along the line joining their centers, and that depend in some way on their distance apart. These are the so-called *central interactions.*

Frictional forces are examples of non-conservative forces.

There are two other major methods of solving dynamical problems that differ in fundamental ways from the method of Newtonian dynamics; they are Lagrangian dynamics and Hamiltonian dynamics. We shall delay a discussion of these more general methods until our study of the Calculus of Variations in Chapter **9**.

4.6 Particle interactions

4.6.1 Elastic collisions

Studies of the collisions among objects, first made in the 17th-century, led to the discovery of two basic laws of Nature: the conservation of linear momentum, and the conservation of kinetic energy associated with a special class of collisions called *elastic collisions.*

The conservation of linear momentum in an isolated system forms the basis for a quantitative discussion of all problems that involve the interactions between particles. The present discussion will be limited to an analysis of the elastic collision between two particles. A typical two-body collision, in which an object of mass m_1 and momentum $\mathbf{p}_1$ makes a grazing collision with another object of mass m_2 and momentum $\mathbf{p}_2$ ($p_2 < p_1$), is shown in the following diagram. (The coordinates are chosen so that the vectors $\mathbf{p}_1$ and $\mathbf{p}_2$ have the same directions). After the collision, the two objects move in directions characterized by the angles θ and ϕ with momenta $\mathbf{p}_1'$ and $\mathbf{p}_2'$.

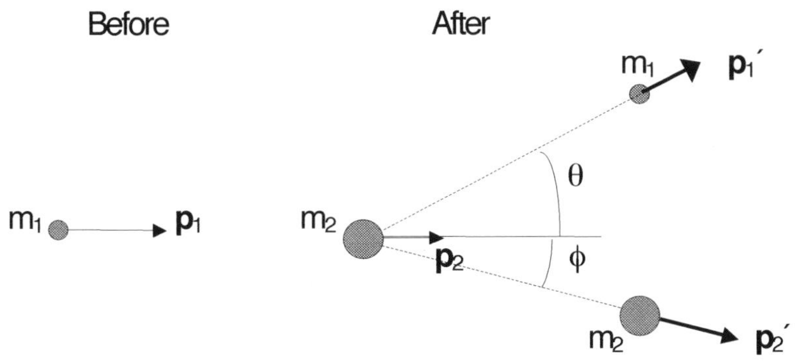

If there are no external forces acting on the particles so that the changes in their states of motion come about as a result of their mutual interactions alone, the total linear momentum of the system is conserved. We therefore have

$$\mathbf{p}_1 + \mathbf{p}_2 = \mathbf{p}_1' + \mathbf{p}_2' \tag{4.33}$$

or, rearranging to give the momentum transfer,

$$\mathbf{p}_1 - \mathbf{p}_1' = \mathbf{p}_2' - \mathbf{p}_2 .$$

The kinetic energy of a particle, T is related to the square of its momentum ($T = p^2/2m$); we therefore form the scalar product of the vector equation for the momentum transfer, to obtain

$$p_1^2 - 2\mathbf{p}_1 \cdot \mathbf{p}_1' + p_1'^2 = p_2'^2 - 2\mathbf{p}_2' \cdot \mathbf{p}_2 + p_2^2 . \tag{4.34}$$

Introducing the scattering angles θ and ϕ, we have

$$p_1^2 - 2p_1p_1'\cos\theta + p_1'^2 = p_2^2 - 2p_2p_2'\cos\phi + p_2'^2 .$$

This equation can be written

$$p_1'^2(x^2 - 2x\cos\theta + 1) = p_2'^2(y^2 - 2y\cos\phi + 1) \quad (4.35)$$

where

$$x = p_1/p_1' \text{ and } y = p_2/p_2' .$$

If we choose a frame in which $\mathbf{p}_2 = 0$ then $y = 0$ and we have

$$x^2 - 2x\cos\theta + 1 = (p_2'/p_1')^2 . \quad (4.36)$$

If the collision is elastic, the kinetic energy of the system is conserved, so that

$$T_1 + 0 = T_1' + T_2' \; (T_2 = 0 \text{ because } \mathbf{p}_2 = 0). \quad (4.37)$$

Substituting $T_i = p_i^2/2m_i$, and rearranging, gives

$$(p_2'/p_1')^2 = (m_2/m_1)(x^2 - 1) .$$

We therefore obtain a quadratic equation in x:

$$x^2 + 2x(m_1/(m_2 - m_1))\cos\theta - [(m_2 + m_1)/(m_2 - m_1)] = 0 .$$

The valid solution of this equation is

$$x = (T_1/T_1')^{1/2} = -(m_1/(m_2 - m_1))\cos\theta$$

$$+ \{(m_1/(m_2 - m_1))^2\cos^2\theta + [(m_2 + m_1)/(m_2 - m_1)]\}^{1/2}. \quad (4.38)$$

If $m_1 = m_2$, the solution is $x = 1/\cos\theta$, in which case

$$T_1' = T_1\cos^2\theta . \quad (4.39)$$

In the frame in which $\mathbf{p}_2 = 0$, a geometrical analysis of the two-body collision is useful. We have

$$\mathbf{p}_1 + (-\mathbf{p}_1') = \mathbf{p}_2', \quad (4.40)$$

leading to

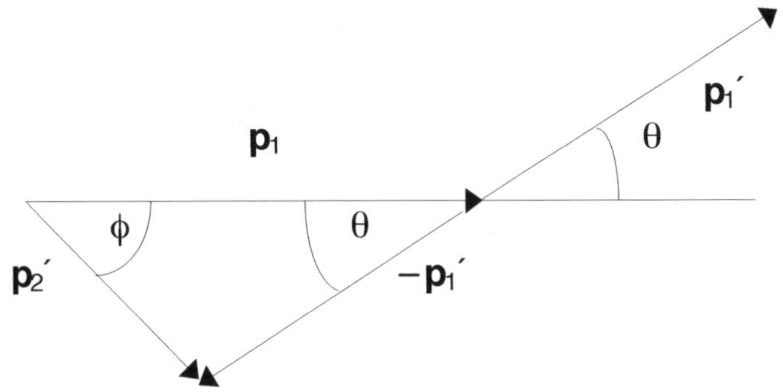

If the masses are equal then

$$p_1' = p_1 \cos\theta .$$

In this case, the two particles always emerge from the elastic collision at right angles to each other ($\theta + \phi = 90^o$).

In the early 1930's, the measured angle between two outgoing high-speed nuclear particles of equal mass was shown to differ from 90°. Such experiments clearly demonstrated the breakdown of Newtonian dynamics in these interactions.

4.6.2 Inelastic collisions

Collisions between everyday objects are never perfectly elastic. An object that has an internal structure can undergo inelastic collisions involving changes in its structure. Inelastic collisions are found to obey two laws; they are

1) the conservation of linear momentum

and

2) an empirical law, due to Newton, that states that the relative velocity of the colliding objects, measured along their line of centers immediately after impact, is $-e$ times their relative velocity before impact.

The quantity e is called the coefficient of restitution. Its value depends on the nature of the materials of the colliding objects. For very hard substances such as steel, e is close to unity, whereas for very soft materials such as putty, e approaches zero.

Consider, in the simplest case, the impact of two deformable spheres with masses m_1 and m_2. Let their velocities be v_1 and v_2, and v_1' and v_2' (along their line of centers) before and after impact, respectively. The linear momentum is conserved, therefore

$$m_1v_1 + m_2v_2 = m_1v_1' + m_2v_2'$$

and, using Newton's empirical law,

$$v_1' - v_2' = -e(v_1 - v_2)\,. \qquad (4.41)$$

Rearranging these equations we can obtain the values v_1' and v_2' , after impact, in terms of their values before impact:

$$v_1' = [m_1v_1 + m_2v_2 - em_2(v_1 - v_2)]/(m_1 + m_2), \qquad (4.42)$$

and

$$v_2' = [m_1v_1 + m_2v_2 + em_1(v_1 - v_2)]/(m_1 + m_2)\,. \qquad (4.43)$$

If the two spheres initially move in directions that are not co-linear, the above method of analysis is still valid because the momenta can be resolved into components along and perpendicular to a chosen axis. The perpendicular components remain unchanged by the impact.

We shall find that the classical approach to a quantitative study of inelastic collisions must be radically altered when we treat the subject within the framework of Special Relativity. It will be shown that the combined mass $(m_1 + m_2)$ of the colliding objects is *not conserved* in an inelastic collision.

4.7 The motion of rigid bodies

Newton's Laws of Motion apply to every point-like mass in an object of finite size. The smallest objects of practical size contain very large numbers of microscopic particles — Avogadro's number is about 6 $\times 10^{23}$ atoms per gram-atom. The motions of the individual microscopic particles in an extended object can be analyzed in terms of the motion of their equivalent total mass, located at the *center of mass* of the object.

4.7.1 The center of mass

For a system of discrete masses m_i, located at the vector positions $\mathbf{r}_i$, the position $\mathbf{r}_{CM}$ of the center of mass is defined as

$$\mathbf{r}_{CM} \equiv \sum_i m_i\mathbf{r}_i / \sum_i m_i = \sum_i m_i\mathbf{r}_i / M, \text{ where M is the total mass.} \quad (4.44)$$

The center of mass (CM) of an (idealized) continuous distribution of mass with a density ρ (mass/volume), can be obtained by considering an element of volume dV with an elemental mass dm. We then have

$$dm = \rho dV. \quad (4.45)$$

The position of the CM is therefore

$$\mathbf{r}_{CM} = (1/M)\int \mathbf{r}dm = (1/M)\int \mathbf{r}\rho dV. \quad (4.46)$$

The Cartesian components of $\mathbf{r}_{CM}$ are

$$x_{iCM} = (1/M)\int x_i\rho dV. \quad (4.47)$$

In non-uniform materials, the density is a function of r.

4.7.2 Kinetic energy of a rigid body in general motion

Consider a rigid body that has both translational and rotational motion in a plane. Let the angular velocity ω be constant. At an arbitrary time t, we have

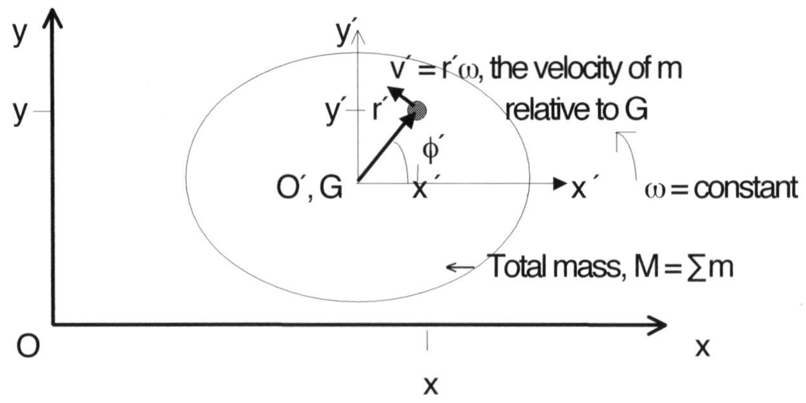

Let the coordinates of an element of mass m of the body be [x, y] in the fixed frame (origin O) and [x´, y´] in the frame moving with the center of mass, G (origin O´), and let u and v be the components of velocity of G, in the fixed frame. For constant angular velocity ω, the instantaneous velocity of the element of mass m, relative to G has a direction perpendicular to the radius vector **r**´, and a magnitude

$$v' = r'\omega. \tag{4.48}$$

The components of the instantaneous velocity of G, relative to the fixed frame, are

u in the x-direction, and

v in the y-direction.

The velocity components of m in the [x, y]-frame are therefore

$$u - r'\omega\sin\phi' = u - y'\omega \text{ in the x-direction,}$$

and

$$v + r'\omega\cos\phi' = v + x'\omega \text{ in the y-direction.}$$

The kinetic energy of the body, E_K, of mass M is therefore

$$E_K = (1/2)\sum m\{(u - y'\omega)^2 + (v + x'\omega)^2\} \tag{4.49}$$

$$= (1/2)M(u^2 + v^2) + (1/2)\omega^2\sum m(x'^2 + y'^2)$$

$$- u\omega\sum my' + v\omega\sum mx'.$$

Therefore

$$E_K = (1/2)Mv_G^2 + (1/2)I_G\omega^2, \tag{4.50}$$

where

$v_G = (u^2 + v^2)^{1/2}$ the speed of G, relative to the fixed frame,

$\sum my' = \sum mx' = 0$, by definition of the center of mass,

and

$I_G = \sum m(x'^2 + y'^2) = \sum mr'^2$, is called the *moment of imertia* of M about an axis through G, perpendicular to the plane.

We see that the total kinetic energy of the moving object of mass M is made up of two parts,

1) the kinetic energy of *translation* of the whole mass moving with the velocity of the center of mass ,

and

2) the kinetic energy of *rotation* of the whole mass about its center of mass.

4.8 Angular velocity and the instantaneous center of rotation

The angular velocity of a body is defined as the rate of increase of the angle between any line AB, fixed in the body, and any line fixed in the plane of the motion. If ϕ is the instantaneous angle between AB and an axis Oy, in the plane, then the angular velocity is $d\phi/dt$.

Consider a circular disc of radius a, that rolls without sliding in contact with a line Ox, and let ϕ be the instantaneous angle that the fixed line AB in the disc makes with the y-axis. At t = 0, the rolling begins with the point B touching the origin, O:

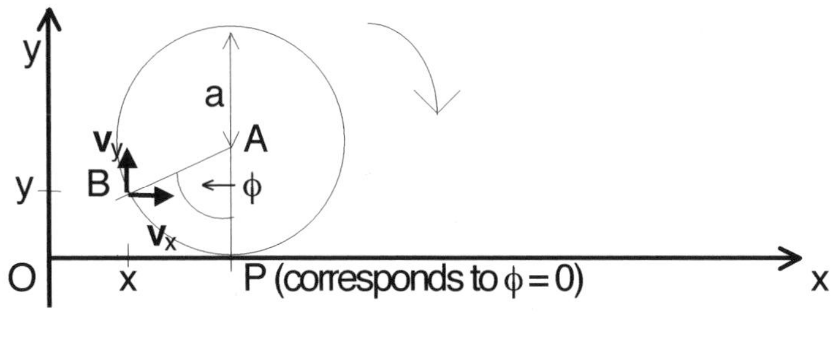

At time t, after the rolling begins, the coordinates of B[x, y] are

$$x = OP - a\sin\phi = BP - a\sin\phi = a\phi - a\sin\phi = a(\phi - \sin\phi),$$

and

$$y = AP - a\cos\phi = a(1 - \cos\phi).$$

The components of the velocity of B are therefore

$$v_x = dx/dt = a(d\phi/dt)(1 - \cos\phi), \tag{4.51}$$

and

$$v_y = dy/dt = a(d\phi/dt)\sin\phi. \tag{4.52}$$

The components of the acceleration of B are

$$a_x = dv_x/dt = (d/dt)(a(d\phi/dt)(1 - \cos\phi)) \tag{4.53}$$

$$= a(d\phi/dt)^2\sin\phi + a(1 - \cos\phi)(d^2\phi/dt^2),$$

and

$$a_y = dv_y/dt = (d/dt)(a(d\phi/dt)\sin\phi) \tag{4.54}$$

$$= a(d\phi/dt)^2\cos\phi + a\sin\phi(d^2\phi/dt^2).$$

If $\phi = 0$,

dx/dt = 0 and dy/dt = 0, which means that the point P has no instantaneous velocity. The point B is therefore instantaneously rotating about P with a velocity equal to $2a\sin(\phi/2)(d\phi/dt)$; P is a "center of rotation".

Also,

$d^2x/dt^2 = 0$ and $d^2y/dt^2 = a(d\phi/dt)^2$, the point of contact only has an acceleration towards the center.

4.9 An application of the Newtonian method

The following example illustrates the use of some basic principles of classical dynamics, such as the conservation of linear momentum, the conservation of energy, and instantaneous rotation about a moving point:

Consider a perfectly smooth, straight horizontal rod with a ring of mass M that can slide along the rod. Attached to the ring is a straight, hinged rod of length L and of negligible mass; it has a mass m at its end. At time t = 0, the system is held in a horizontal position in the constant gravitational field of the Earth.

At t = 0:

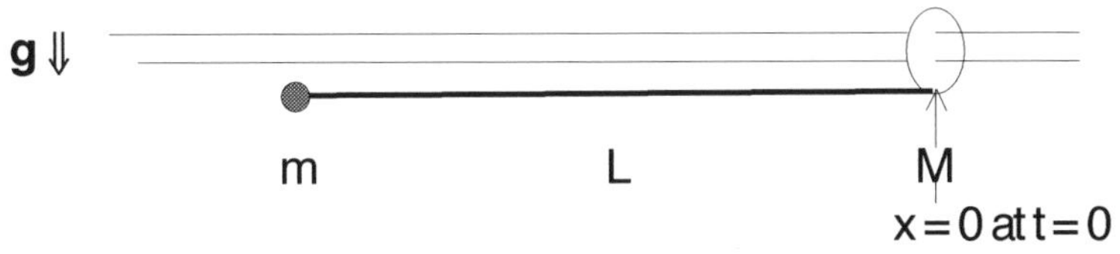

At t = 0, the mass m is released and falls under gravity. At time t, we have

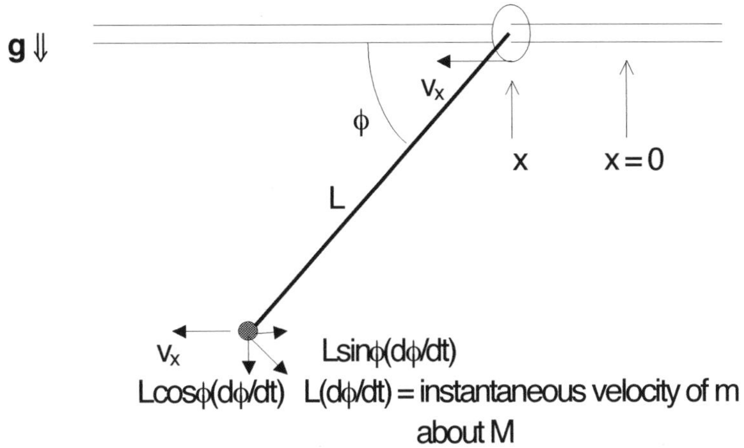

There are no external forces acting on the system in the x-direction and therefore the horizontal momentum remains zero:

$$M(dx/dt) + m((dx/dt) - L\sin\phi(d\phi/dt)) = 0. \quad (4.56)$$

Integrating, we have

$$Mx + mx + mL\cos\phi = \text{constant}. \quad (4.57)$$

If $x = 0$ and $\phi = 0$ at $t = 0$, then

$$mL = \text{constant}, \tag{4.58}$$

therefore

$$(M + m)x + mL(\cos\phi - 1) = 0,$$

so that

$$x = mL(1 - \cos\phi)/(M + m). \tag{4.59}$$

We see that the instantaneous position x(t) is obtained by integrating the momentum equation.

The equation of conservation of energy can now be used; it is

$$(M/2)v_x^2 + (m/2)(v_x - L\sin\phi(d\phi/dt))^2 + (m/2)(L\cos\phi(d\phi/dt))^2 = mgL\sin\phi.$$

(The change in kinetic energy is equal to the change in the potential energy).

Rearranging, gives

$$(M + m)v_x^2 - 2mL\sin\phi v_x(d\phi/dt) + (mL^2(d\phi/dt)^2 - 2mgL\sin\phi) = 0. \tag{4.60}$$

This is a quadratic in v_x with a solution

$$(M + m)v_x = mL\sin\phi(d\phi/dt)[1 \pm \{1 - [(M + m)(mL^2(d\phi/dt)^2$$

$$- 2mLg\sin\phi)]/[m^2L^2(d\phi/dt)^2\sin^2\phi]\}^{1/2}].$$

The left-hand side of this equation is also given by the momentum equation:

$$(M + m)v_x = mL\sin\phi(d\phi/dt).$$

We therefore obtain, after substitution and rearrangement,

$$d\phi/dt = \{[2(M + m)g\sin\phi]/[L(M + m\cos^2\phi)]\}^{1/2}, \tag{4.61}$$

the angular velocity of the rod of length L at time t.

PROBLEMS

4-1 A straight uniform rod of mass m and length 2l is held at an angle θ_0 to the vertical. Its lower end rests on a perfectly smooth horizontal surface. The rod is released and falls under gravity. At time t after the motion begins, we have

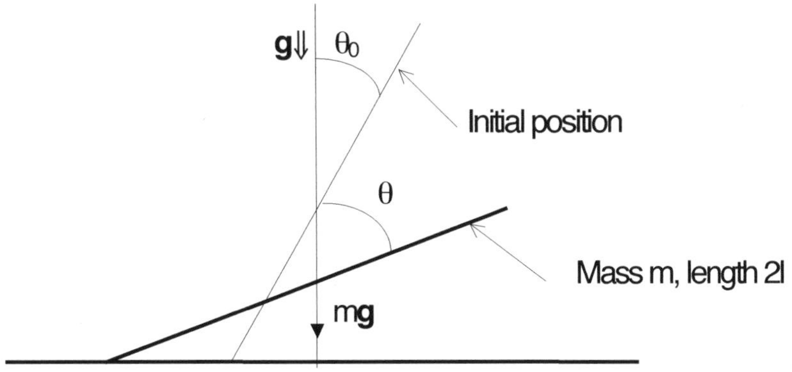

If the moment of inertia of the rod about an axis through its center of mass, perpendicular to the plane of the motion, is $ml^2/3$, prove that the angular velocity of the rod when it makes an angle θ with the vertical, is

$$d\theta/dt = \{6g(\cos\theta_0 - \cos\theta)/l(1 + 3\sin^2\theta)\}^{1/2}.$$

4-2 Show that the center of mass of a uniform solid hemisphere of radius R is 3R/8 above the center of its plane surface.

4-3 Show that the moment of inertia of a uniform solid sphere of radius R and mass M about a diameter is $2MR^2/5$.

4-4 A uniform solid sphere of mass m and radius r can roll, under gravity, on the inner surface of a perfectly rough spherical surface of radius R. The motion is in a vertical plane. At time t during the motion, we have

Rolling sphere, mass m and radius r

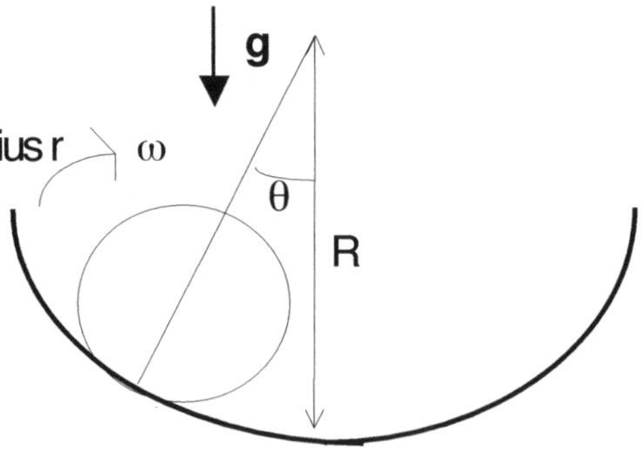

Show that

$$d^2\theta/dt^2 + [5g/(7(R-r))]\sin\theta = 0.$$

As a preliminary result, show that $r\omega = (R-r)(d\theta/dt)$ for rolling motion without slipping.

4-5 A particle of mass m hangs on an inextensible string of length l and negligible mass. The string is attached to a fixed point O. The mass oscillates in a vertical plane under gravity. At time t, we have

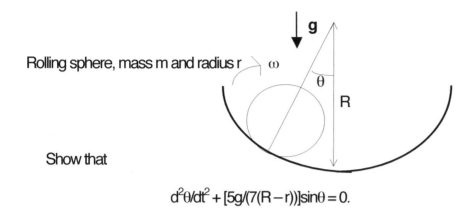

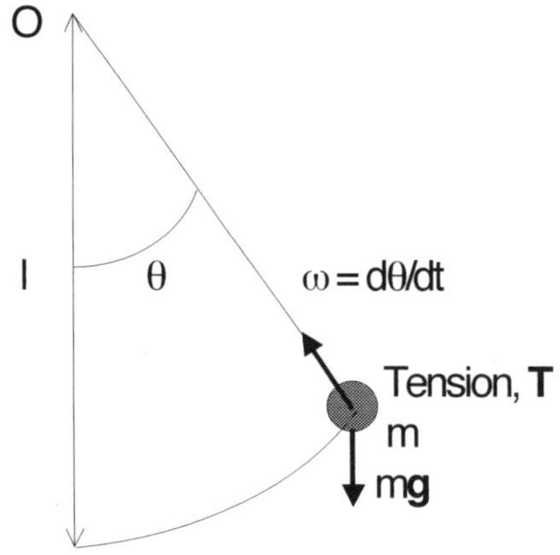

Show that

1) $d^2\theta/dt^2 + (g/l)\sin\theta = 0.$

2) $\omega^2 = (2g/l)[\cos\theta - \cos\theta_0]$, where θ_0 is the initial angle of the string with respect to the vertical, so that $\omega = 0$ when $\theta = \theta_0$. This equation gives the angular velocity in any position.

4-6 Let l_0 be the natural length of an elastic string fixed at the point O. The string has a negligible mass. Let a mass m be attached to the string, and let it stretch the string until the equilibrium position is reached. The tension in the string is given by Hooke's law:

Tension, T = λ (extension)/original length, where λ is a constant for a given material.

The mass is displaced vertically from its equilibrium position, and oscillates under gravity. We have

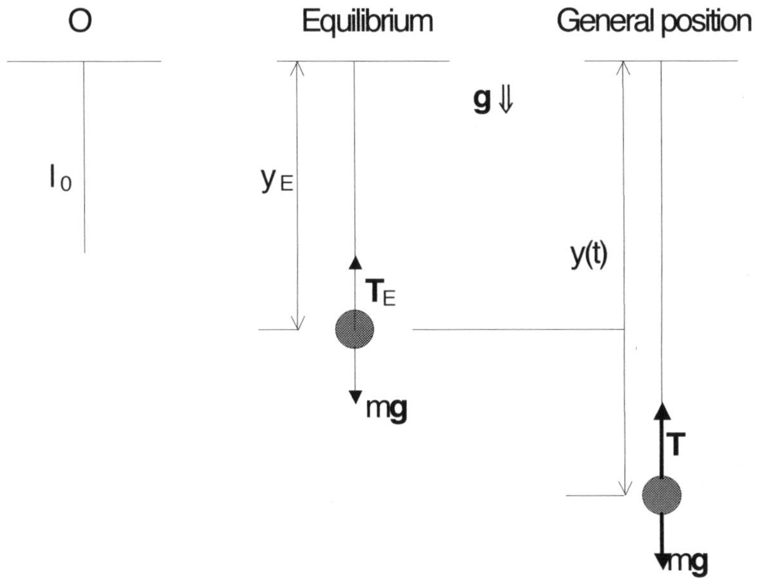

Show that the mass oscillates about the equilibrium position with simple harmonic motion, and that

$y(t) = l_0 + (mgl_0/\lambda)\{1 - \cos[t\sqrt{\lambda/ml_0}]\}$ (starts with zero velocity at $y(0) = l_0$)

4-7 A dynamical system is in stable equilibrium if the system tends to return to its original state if slightly displaced. A system is in a position of equilibrium when the height of its center of gravity is a maximum or a minimum. Consider a rod of mass m with one end resting on a perfectly smooth vertical wall OA and the other end on a perfectly smooth inclined plane, OB. Show that, in the position of equilibrium

$\cot\theta = 2\tan\phi$, where the angles are given in the diagram:

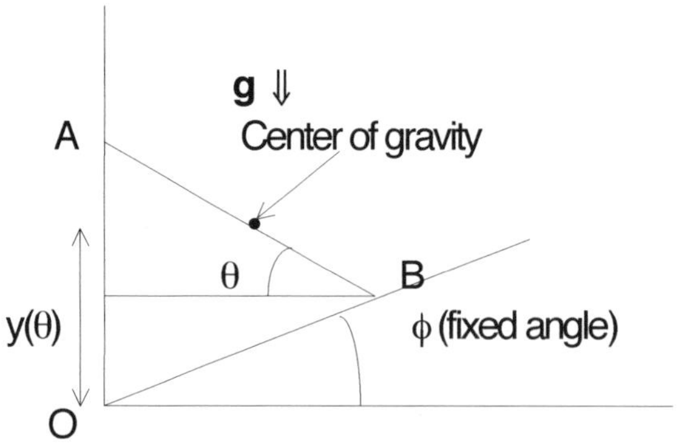

Find $y = f(\theta)$, and show, by considering derivatives, that this is a state of unstable equilibrium.

4-8 A particle A of mass $m_A = 1$ unit, scatters elastically from a stationary particle B of mass $m_B = 2$ units. If A scatters through an angle θ, show that the ratio of the kinetic energies of A, before (T_A) and after (T_A') scattering is

$$(T_A/T_A') = (-\cos\theta + \sqrt{3 + \cos^2\theta})^2.$$

Sketch the form of the variation of this ratio with angle in the range $0 \le \theta \le \pi$.

(This problem is met in practice in low-energy neutron-deuteron scattering).

5

INVARIANCE PRINCIPLES AND CONSERVATION LAWS

5.1 Invariance of the potential under translations and the conservation of linear momentum

The equation of motion of a Newtonian particle of mass m moving along the x-axis under the influence of a force F_x is

$$md^2x/dt^2 = F_x . \tag{5.1}$$

If F_x can be represented by a potential V(x) then

$$md^2x/dt^2 = -dV(x)/dx . \tag{5.2}$$

In the special case in which the potential is not a function of x, the equation of motion becomes

$$md^2x/dt^2 = 0,$$

or

$$md(v_x)/dt = 0. \tag{5.3}$$

Integrating this equation gives

$$mv_x = \text{constant}. \tag{5.4}$$

We see that the linear momentum of the particle is constant if the potential is independent of the position of the particle.

5.2 Invariance of the potential under rotations and the conservation of angular momentum

Let a Newtonian particle of mass m move in the plane about a fixed origin, O, under the influence of a force **F**. The equations of motion, in the x-and y-directions, are

$$md^2x/dt^2 = F_x \text{ and } md^2y/dt^2 = F_y. \tag{5.5 a, b}$$

If the force can be represented by a potential V(x, y) then we can write

$$md^2x/dt^2 = -\partial V/\partial x \text{ and } md^2y/dt^2 = -\partial V/\partial y\,. \tag{5.6 a, b}$$

The total differential of the potential is

$$dV = (\partial V/\partial x)dx + (\partial V/\partial y)dy.$$

Let a transformation from Cartesian to polar coordinates be made using the standard linear equations

$$x = r\cos\phi \text{ and } y = r\sin\phi\,.$$

The partial derivatives are

$$\partial x/\partial\phi = -r\sin\phi = -y,\ \partial x/\partial r = \cos\phi,\ \partial y/\partial\phi = r\cos\phi = x, \text{ and } \partial y/\partial r = \sin\phi\,.$$

We therefore have

$$\partial V/\partial\phi = (\partial V/\partial x)(\partial x/\partial\phi) + (\partial V/\partial y)(\partial y/\partial\phi) \tag{5.7}$$

$$= (\partial V/\partial x)(-y) + (\partial V/\partial y)(x)$$

$$= yF_x + x(-F_y)$$

$$= m(ya_x - xa_y) \text{ (a_x and a_y are the components of acceleration)}$$

$$= m(d/dt)(yv_x - xv_y) \text{ (v_x and v_y are the components of velocity).}$$

If the potential is independent of the angle ϕ then

$$\partial V/\partial\phi = 0, \tag{5.8}$$

in which case

$$m(d/dt)(yv_x - xv_y) = 0$$

and therefore

$$m(yv_x - xv_y) = \text{a constant.} \tag{5.9}$$

The quantity on the left-hand side of this equation is the angular momentum ($yp_x - xp_y$) of the mass about the fixed origin. We therefore see that if the potential is invariant under rotations about the origin (independent of the angle ϕ), the angular momentum of the mass about the origin is conserved.

In Chapter **9**, we shall treat the subject of invariance principles and conservation laws in a more general way, using arguments that involve the Lagrangians and Hamiltonians of dynamical systems.

6

EINSTEINIAN DYNAMICS

6.1 4-momentum and the energy-momentum invariant

In Classical Mechanics, the concept of momentum is important because of its rôle as an invariant in an isolated system. We therefore introduce the concept of 4-momentum in Relativistic Mechanics in order to find possible Lorentz invariants involving this new quantity. The contravariant 4-momentum is defined as:

$$P^{\mu} = mV^{\mu} \tag{6.1}$$

where m is the mass of the particle. (It is a Lorentz scalar — the mass measured in the rest frame of the particle).

The scalar product is

$$P^{\mu}P_{\mu} = (mc)^2. \tag{6.2}$$

Now,

$$P^{\mu} = [m\gamma c, m\gamma \mathbf{v}_N] \tag{6.3}$$

therefore,

$$P^{\mu}P_{\mu} = (m\gamma c)^2 - (m\gamma \mathbf{v}_N)^2.$$

Writing

$M = \gamma m$, the relativistic mass, we obtain

$$P^{\mu}P_{\mu} = (Mc)^2 - (M\mathbf{v}_N)^2 = (mc)^2. \tag{6.4}$$

Multiplying throughout by c^2 gives

$$M^2c^4 - M^2{v_N}^2c^2 = m^2c^4. \tag{6.5}$$

The quantity Mc^2 has dimensions of energy; we therefore write

$$E = Mc^2, \tag{6.6}$$

the total energy of a freely moving particle.

This leads to the fundamental invariant of dynamics

$$c^2P^{\mu}P_{\mu} = E^2 - (\mathbf{p}c)^2 = E^{o2} \tag{6.7}$$

where

$E^o = mc^2$ is the rest energy of the particle, and $\mathbf{p}$ is its *relativistic 3-momentum.*

The total energy can be written:

$$E = \gamma E^o = E^o + T, \tag{6.8}$$

where

$$T = E^o(\gamma - 1), \tag{6.9}$$

the *relativistic kinetic energy.*

The magnitude of the 4-momentum is a Lorentz invariant

$$|P^{\mu}| = mc. \tag{6.10}$$

The 4- momentum transforms as follows:

$$P'^{\mu} = \mathbf{L}P^{\mu}. \tag{6.11}$$

6.2 The relativistic Doppler shift

For relative motion along the x-axis, the equation $P'^{\mu} = \mathbf{L}P^{\mu}$ is equivalent to the equations

$$E' = \gamma E - \beta\gamma cp^x \tag{6.12}$$

and,

$$cp'^x = -\beta\gamma E + \gamma cp^x. \tag{6.13}$$

Using the Planck-Einstein equations $E = h\nu$ and $E = p^x c$ for photons, the energy equation becomes

$$\nu' = \gamma\nu - \beta\gamma\nu$$

$$= \gamma\nu(1 - \beta)$$

$$= \nu(1-\beta)/(1-\beta^2)^{1/2}$$

$$= \nu\{(1-\beta)/(1+\beta)\}^{1/2}. \qquad (6.14)$$

This is the relativistic Doppler shift for photons of the frequency ν', measured in an inertial frame (primed) in terms of the frequency ν measured in another inertial frame.

6.3 Relativistic collisions and the conservation of 4-momentum

Consider the interaction between two particles, 1 and 2, to form two particles, 3 and 4. (3 and 4 are not necessarily the same as 1 and 2). The contravariant 4-momenta are P_i^{μ} :

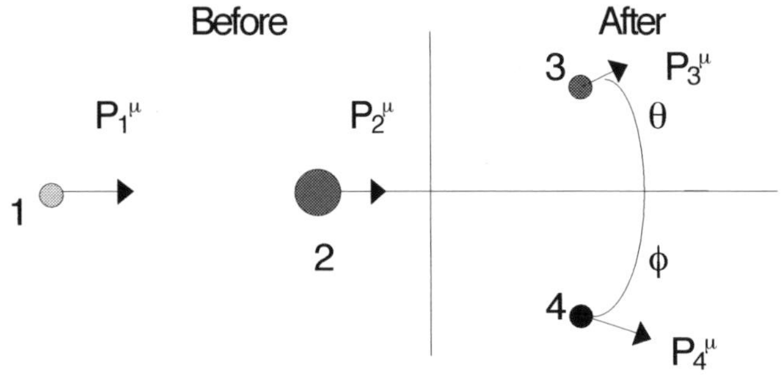

$$1 + 2 \rightarrow 3 + 4$$

All experiments are consistent with the fact that the 4-momentum of the system is conserved. We have, for the contravariant 4-momentum vectors of the interacting particles,

$$P_1^{\mu} + P_2^{\mu} = P_3^{\mu} + P_4^{\mu} \qquad (6.15)$$

initial "free" state (↑ under $P_1^{\mu} + P_2^{\mu}$) final "free" state (↑ under $P_3^{\mu} + P_4^{\mu}$)

and a similar equation for the covariant 4-momentum vectors,

$$P_{1_\mu} + P_{2_\mu} = P_{3_\mu} + P_{4_\mu}. \qquad (6.16)$$

If we are interested in the change $P_1^{\mu} \rightarrow P_3^{\mu}$, then we require

$$P_1^{\mu} - P_3^{\mu} = P_4^{\mu} - P_2^{\mu} \qquad (6.17)$$

and

$$P_{1\mu} - P_{3\mu} = P_{4\mu} - P_{2\mu}. \tag{6.18}$$

Forming the invariant scalar products, and using $P_{i\mu}P_i^{\mu} = (E_i^0/c)^2$, we obtain

$$(E_1^0/c)^2 - 2(E_1E_3/c^2 - \mathbf{p}_1\cdot\mathbf{p}_3) + (E_3^0/c)^2$$

$$= (E_4^0/c)^2 - 2(E_2E_4/c^2 - \mathbf{p}_2\cdot\mathbf{p}_4) + (E_2^0/c)^2. \tag{6.19}$$

Introducing the scattering angles, θ and ϕ, this equation becomes

$$E_1^{02} - 2(E_1E_3 - c^2p_1p_3\cos\theta) + E_3^{02} = E_2^{02} - 2(E_2E_4 - c^2p_2p_4\cos\phi) + E_4^{02}.$$

If we choose a reference frame in which particle 2 is at rest (the LAB frame), then $\mathbf{p}_2 = 0$ and $E_2 = E_2^0$, so that

$$E_1^{02} - 2(E_1E_3 - c^2p_1p_3\cos\theta) + E_3^{02} = E_2^{02} - 2E_2^0E_4 + E_4^{02}. \tag{6.20}$$

The total energy of the system is conserved, therefore

$$E_1 + E_2 = E_3 + E_4 = E_1 + E_2^0 \tag{6.21}$$

or

$$E_4 = E_1 + E_2^0 - E_3$$

Eliminating E_4 from the above "scalar product' equation gives

$$E_1^{02} - 2(E_1E_3 - c^2p_1p_3\cos\theta) + E_3^{02} = E_4^{02} - E_2^{02} - 2E_2^0(E_1 - E_3). \tag{6.22}$$

This is the basic equation for all interactions in which two relativistic entities in the initial state interact to give two relativistic entities in the final state. It applies equally well to interactions that involve massive and massless entities.

6.3.1 The Compton effect

The general method discussed in the previous section can be used to provide an exact analysis of Compton's famous experiment in which the scattering of a photon by a stationary, free electron was studied. In this example, we have

$E_1 = E_{ph}$ (the incident photon energy), $E_2 = E_e^0$ (the rest energy of the stationary electron, the "target"), $E_3 = E_{ph}'$ (the energy of the scattered photon), and $E_4 = E_e'$ (the energy of the recoiling electron). The "rest energy" of the photon is zero:

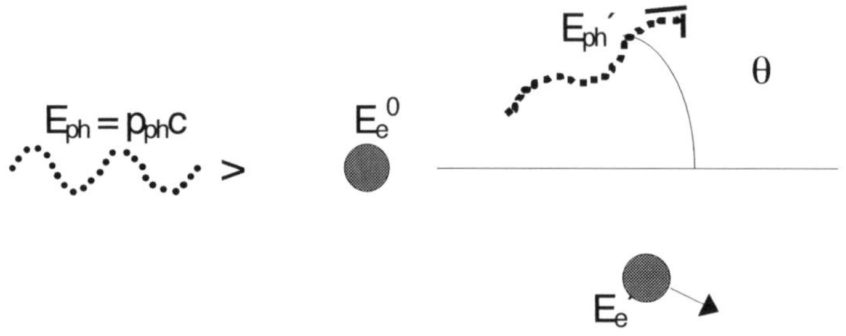

The general equation (6.22), is now

$$0 - 2(E_{ph}E_{ph}' - E_{ph}E_{ph}'\cos\theta) = E_e^{0\,2} - 2E_e^0(E_{ph} + E_e^0 - E_{ph}') + E_e^{0\,2} \tag{6.23}$$

or

$$-2E_{ph}E_{ph}'(1 - \cos\theta) = -2E_e^0(E_{ph} - E_{ph}')$$

so that

$$E_{ph} - E_{ph}' = E_{ph}E_{ph}'(1 - \cos\theta)/E_e^0\,. \tag{6.24}$$

Compton measured the energy-loss of the photon on scattering and its $\cos\theta$ - dependence.

6.4 Relativistic inelastic collisions

We shall consider an inelastic collision between a particle 1 and a particle 2 (initially at rest) to form a composite particle 3. In such a collision, the 4-momentum is conserved (as it is in an elastic collision) however, the kinetic energy is not conserved. Part of the kinetic energy of particle 1 is transformed into excitation energy of the composite particle 3. This excitation energy can take many forms — heat energy, rotational energy, and the excitation of quantum states at the microscopic level.

The inelastic collision is as shown:

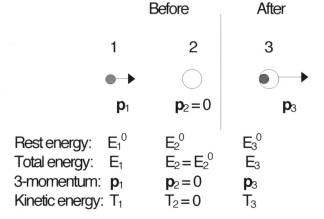

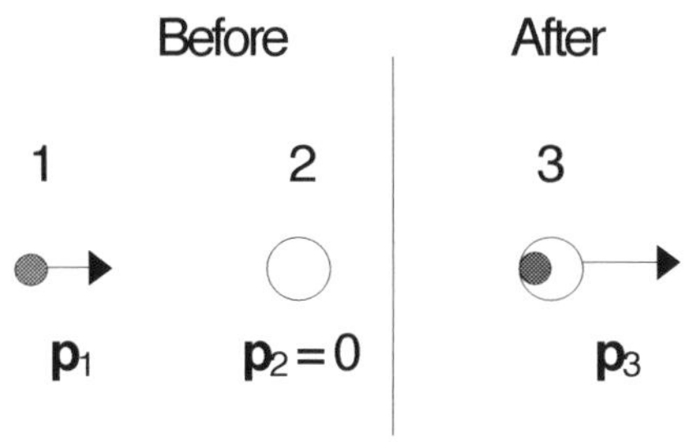

Rest energy:	$E_1^{\,0}$	$E_2^{\,0}$	$E_3^{\,0}$
Total energy:	E_1	$E_2 = E_2^{\,0}$	E_3
3-momentum:	$\mathbf{p}_1$	$\mathbf{p}_2 = 0$	$\mathbf{p}_3$
Kinetic energy:	T_1	$T_2 = 0$	T_3

In this problem, we shall use the energy-momentum invariants associated with each particle, directly:

i) $E_1^{\,2} - (\mathbf{p}_1 c)^2 = E_1^{\,0\,2}$ (6.25)

ii) $E_2^{\,2} \qquad\qquad = E_2^{\,0\,2}$ (6.26)

iii) $E_3^{\,2} - (\mathbf{p}_3 c)^2 = E_3^{\,0\,2}$. (6.27)

The total energy is conserved, therefore

$$E_1 + E_2 = E_3 = E_1 + E_2^{\,0}. \tag{6.28}$$

Introducing the kinetic energies of the particles, we have

$$(T_1 + E_1^{\,0}) + E_2^{\,0} = T_3 + E_3^{\,0}. \tag{6.29}$$

The 3-momentum is conserved, therefore

$$\mathbf{p}_1 + 0 = \mathbf{p}_3. \tag{6.30}$$

Using

$$E_3^{\,0\,2} = E_3^{\,2} - (\mathbf{p}_3 c)^2, \tag{6.31}$$

we obtain

$$E_3^{0\,2} = (E_1 + E_2^0)^2 - (\mathbf{p}_3c)^2$$

$$= E_1^2 + 2E_1E_2^0 + E_2^{0\,2} - (\mathbf{p}_1c)^2$$

$$= E_1^{0\,2} + 2E_1E_2^0 + E_2^{0\,2}$$

$$= E_1^{0\,2} + E_2^{0\,2} + 2(T_1 + E_1^0)E_2^0 \qquad (6.32)$$

or

$$E_3^{0\,2} = (E_1^0 + E_2^0)^2 + 2T_1E_2^0 \quad (E_3^0 > E_1^0 + E_2^0). \qquad (6.33)$$

Using $T_1 = \gamma_1E_1^0 - E_1^0$, where $\gamma_1 = (1 - \beta_1^2)^{-1/2}$ and $\beta_1 = v_1/c$, we have

$$E_3^{0\,2} = E_1^{0\,2} + E_2^{0\,2} + 2\gamma_1E_1^0E_2^0. \qquad (6.34)$$

If two identical particles make a completely inelastic collision then

$$E_3^{0\,2} = 2(\gamma_1 + 1)E_1^{0\,2}. \qquad (6.35)$$

6.5 The Mandelstam variables

In discussions of relativistic interactions it is often useful to introduce additional Lorentz invariants that are known as Mandelstam variables. They are, for the special case of two particles in the initial and final states ($1 + 2 \rightarrow 3 + 4$):

$$s = (P_1^\mu + P_2^\mu)[P_{1\mu} + P_{2\mu}], \text{ the total 4-momentum invariant}$$

$$= ((E_1 + E_2)/c, (\mathbf{p}_1 + \mathbf{p}_2))[(E_1 + E_2)/c, -(\mathbf{p}_1 + \mathbf{p}_2)]$$

$$= (E_1 + E_2)^2/c^2 - (\mathbf{p}_1 + \mathbf{p}_2)^2, \text{ a Lorentz invariant}, \qquad (6.36)$$

$$t = (P_1^\mu - P_3^\mu)[P_{1\mu} - P_{3\mu}], \text{ the 4-momentum transfer } (1 \rightarrow 3) \text{ invariant}$$

$$= (E_1 - E_3)^2/c^2 - (\mathbf{p}_1 - \mathbf{p}_3)^2, \text{ a Lorentz invariant}, \qquad (6.37)$$

and

$$u = (P_1^\mu - P_4^\mu)[P_{1\mu} - P_{4\mu}], \text{ the 4-momentum transfer } (1 \rightarrow 4) \text{ invariant}$$

$$= (E_1 - E_4)^2/c^2 - (\mathbf{p}_1 - \mathbf{p}_4)^2, \text{ a Lorentz invariant.} \tag{6.38}$$

Now,

$$sc^2 = E_1^{\,2} + 2E_1E_2 + E_2^{\,2} - (p_1^{\,2} + 2\mathbf{p}_1{\cdot}\mathbf{p}_2 + p_2^{\,2})c^2$$

$$= E_1^{0\,2} + E_2^{0\,2} + 2E_1E_2 - 2\mathbf{p}_1{\cdot}\mathbf{p}_2c^2$$

$$= E_1^{0\,2} + E_2^{0\,2} + 2\underbrace{(E_1, \mathbf{p}_1c)[E_2, -\mathbf{p}_2c]}_{\uparrow\ \text{Lorentz invariant}}. \tag{6.39}$$

The Mandelstam variable sc^2 has the same value in all inertial frames. We therefore evaluate it in the LAB frame, defined by the vectors

$$[E_1^{\,L}, \mathbf{p}_1^{\,L}c] \text{ and } [E_2^{\,L} = E_2^{\,0}, -\mathbf{p}_2^{\,L}c = 0], \tag{6.40}$$

so that

$$2(E_1^{\,L}E_2^{\,L} - \mathbf{p}_1^{\,L}{\cdot}\mathbf{p}_2^{\,L}c^2) = 2E_1^{\,L}E_2^{\,0}, \tag{6.41}$$

and

$$sc^2 = E_1^{0\,2} + E_2^{0\,2} + 2E_1^{\,L}E_2^{\,0}. \tag{6.42}$$

We can evaluate sc^2 in the center-of -mass (CM) frame, defined by the condition

$$\mathbf{p}_1^{\,CM} + \mathbf{p}_2^{\,CM} = 0 \text{ (the total 3-momentum is zero):}$$

$$sc^2 = (E_1^{\,CM} + E_2^{\,CM})^2. \tag{6.43}$$

This is the square of the total CM energy of the system.

6.5.1 The total CM energy and the production of new particles

The quantity $c\sqrt{s}$ is the energy available for the production of new particles, or for exciting the internal structure of particles. We can now obtain the relation between the total CM energy and the LAB energy of the incident particle (1) and the target (2), as follows:

$$sc^2 = E_1^{0\,2} + E_2^{0\,2} + 2E_1^{L}E_2^{0} = (E_1^{CM} + E_2^{CM})^2 = W^2, \text{ say.} \tag{6.44}$$

Here, we have evaluated the left-hand side in the LAB frame, and the right-hand side in the CM frame.

At very high energies, $c\sqrt{s} >> E_1^0$ and E_2^0, the rest energies of the particles in the initial state, in which case,

$$W^2 = sc^2 \approx 2E_2^{L}E_2^{0}. \tag{6.45}$$

The total CM energy, W, available for the production of new particles therefore depends on the square root of the incident laboratory energy. This result led to the development of colliding, or intersecting, beams of particles (such as protons and anti-protons) in order to produce sufficient energy to generate particles with rest masses greater than 100 times the rest mass of the proton ($\approx 10^9$ eV).

6.6 Positron-electron annihilation-in-flight

A discussion of the annihilation-in-flight of a relativistic positron and a stationary electron provides a topical example of the use of relativistic conservation laws. This process, in which two photons are spontaneously generated, has been used as a source of nearly monoenergetic high-energy photons for the study of nuclear photo-disintegration since 1960. The general result for a $1 + 2 \rightarrow 3 + 4$ interaction, given in section **6. 3**, provides the basis for an exact calculation of this process; we have

$E_1 = E_{pos}$ (the incident positron energy), $E_2 = E_e^0$ (the rest energy of the stationary electron), $E_3 = E_{ph1}$ (the energy of the forward-going photon), and $E_4 = E_{ph2}$ (the energy of the backward-going photon). The rest energies of the positron and the electron are equal. The general equation (6.22), now reads

$$E_e^{0 2} - 2\{E_{pos}E_{ph1} - cp_{pos}E_{ph1}(\cos\theta)\} + 0 = 0 - E_e^{0 2} - 2E_e^0(E_{pos} - E_{ph1}) \tag{6.46}$$

therefore,

$$E_{ph1}\{E_{pos} + E_e^0 - [E_{pos}^{\ 2} - E_e^{0 2}]^{1/2}\cos\theta\} = (E_{pos} + E_e^0)E_e^0,$$

giving

$$E_{ph1} = E_e^0/(1 - k\cos\theta) \tag{6.47}$$

where

$$k = [(E_{pos} - E_e^0)/(E_{pos} + E_e^0)]^{1/2}.$$

The maximum energy of the photon, E_{ph1}^{max} occurs when $\theta = 0$, corresponding to motion in the forward direction; its energy is

$$E_{ph1}^{max} = E_{oe}/(1 - k). \tag{6.48}$$

If, for example, the incident total positron energy is 30 MeV, and $E_e^0 = 0.511$MeV then

$$E_{ph1}^{max} = 0.511/[1 - (29.489 / 30.511)^{1/2}] \text{ MeV}$$

$$= 30.25 \text{ MeV}.$$

The forward-going photon has an energy equal to the kinetic energy of the incident positron ($T_1 = 30 - 0.511$ MeV) plus approximately three-quarters of the total rest energy of the positron-electron pair ($2E_e^0 = 1.02$ MeV). Using the conservation of the total energy of the system, we see that the energy of the backward-going photon is approximately 0.25 MeV.

The method of positron-electron annihilation-in-flight provides one of the very few ways of generating nearly monoenergetic photons at high energies.

PROBLEMS

6-1 A particle of rest energy E^0 has a relativistic 3-momentum **p** and a relativistic kinetic energy T. Show that

1) $|\mathbf{p}| = (1/c)(2TE^0)^{1/2}\{1 + (T/2E^0)\}^{1/2}$,

and

2) $|\mathbf{v}| = c\{1 + [E^{02}/T(T + 2E^0)]\}^{-1/2}$, where **v** is the 3-velocity.

6-2 Two similar relativistic particles, A and B, each with rest energy E^0, move towards each other in a straight line. The constant speed of each particle , measured in the LAB frame is $V = \beta c$. Show that their total energy, measured in the rest frame of A, is

$$E^0(1 + \beta^2)/(1 - \beta^2).$$

6-3 An atom of rest energy $E_A{}^0$ is initially at rest. It completely absorbs a photon of energy E_{ph}, and the excited atom of rest energy $E_A{}^{0*}$ recoils freely. If the excitation energy of the atom is given by

$$E_{ex} = E_A{}^{0*} - E_A{}^0, \text{ show that}$$

$$E_{ex} = -E_A{}^0 + E_A{}^0\{1 + (2E_{ph}/E_A{}^0)\}^{1/2}, \text{ exactly.}$$

If, as is often the case, $E_{ph} \ll E_A{}^0$, show that the recoil energy of the atom is

$$E_{recoil} \approx E_{ph}{}^2/2E_A{}^0.$$

Explain how this approximation can be deduced using a Newtonian-like analysis.

6-4 A completely inelastic collision occurs between particle 1 and particle 2 (initially at rest) to form a composite particle, 3. Show that the speed of 3 is

$$v_3 = v_1/\{1 + (E_2{}^0/E_1)\},$$

where v_1 and E_1 are the speed and the total energy of 1, and $E_2{}^0$ is the rest energy of 2.

6-5 Show that the *minimum* energy that a γ-ray must have to just break up a deuteron into a neutron and a proton is $\gamma^{min} \approx 2.23$ MeV, given

$$E_{neut}{}^0 = 939.5656 \text{ MeV,}$$

$$E_{prot}{}^0 = 938.2723 \text{ MeV, and}$$

$$E_{deut}{}^0 = 1875.6134 \text{ MeV.}$$

6-6 In a general relativistic collision:

$$1 + 2 \to \text{n-particles}$$

$\rightarrow (3 + 4 + ...m) + (m+1, m+2. + ...n)$

where the particles $3 \rightarrow m$ are "observed", and the particles $m+1 \rightarrow n$ are "unobserved". We have

$$E_1 + E_2 = (E_3 + E_4 + ...E_m) + (E_{m+1} + E_{m+2} + ...E_n), \text{ the total energy,}$$

$$= E^{obs} + E^{unobs}$$

and

$$\mathbf{p}_1 + \mathbf{p}_2 = \mathbf{p}^{obs} + \mathbf{p}^{unobs}.$$

If W^{unobs}/c^2 is the unobserved (missing) mass of the particles m+1 to n, show that, in the LAB frame

$$(W^{unobs})^2 = (E_1^{\ L} + E_2^{\ 0} - \sum_{[i=3,m]} E_i^{\ L})^2 - (\mathbf{p}_1^{\ L}c - c\sum_{[i=3,m]} \mathbf{p}_i^{\ L})^2.$$

This is the missing (energy)^2 in terms of the observed quantities. This is the principle behind the so-called "missing-mass spectrometers" used in Nuclear and Particle Physics.

6.7 If the contravariant 4-force is defined as $F^\mu = dP^\mu/d\tau = [f^0, \mathbf{f}]$ where τ is the proper time, and P^μ is the contravariant 4-momentum, show that

$F^\mu V_\mu = 0$, where V_μ is the covariant 4-velocity. (The 4- force and the 4-velocity are orthogonal).

Obtain dE/dt in terms of γ, $\mathbf{v}$, and $\mathbf{f}$.

7

NEWTONIAN GRAVITATION

We come now to one of the highlights in the history of intellectual endeavor, namely Newton's Theory of Gravitation. This spectacular work ranks with a handful of masterpieces in Natural Philosophy — the Galileo-Newton Theory of Motion, the Carnot-Clausius-Kelvin Theory of Heat and Thermodynamics, Maxwell's Theory of Electromagnetism, the Maxwell-Boltzmann-Gibbs Theory of Statistical Mechanics, Einstein's Theories of Special and General Relativity, Planck' s Quantum Theory of Radiation, and the Bohr-deBroglie-Schrödinger-Heisenberg Quantum Theory of Matter.

Newton's most significant ideas on Gravitation were developed in his early twenties at a time when the University of Cambridge closed down because of the Great Plague. He returned to his home, a farm at Woolsthorpe – by - Colsterworth in Lincolnshire. It is a part of England dominated by vast, changing skies; a region buffeted by the winds from the North Sea. The thoughts of the young Newton naturally turned skyward — there was little on the ground to stir his imagination except, perhaps, the proverbial apple tree and the faling apple. Newton's work set us on a new course.

Before discussing the details of the theory, it will be useful to give an overview using the simplest model, consistent with logical accuracy. In this way, we can appreciate Newton's radical ideas, and his development of the now standard "Scientific Method " in which a crucial interplay exists between the results of observations and mathematical models that best account for the observations. The great theories are often based upon relatively small numbers of observations. The uncovering of the Laws of Nature requires deep and imaginative thoughts that go far beyond the demonstration of mathematical prowess.

Newton's development of Differential Calculus in the late 1660's was strongly influenced by his attempts to understand, analytically, the empirical ideas concerning motion that had been put forward by

Galileo. In particular, he investigated the analytical properties of motion in curved paths. These properties are required in his Theory of Gravitation. We shall consider motion in 2-dimensions.

7.1 Properties of motion along curved paths in the plane

The velocity of a point in the plane is a vector, drawn at the point, such that its component in any direction is given by the rate of change of the displacement, in that direction. Consider the following diagram

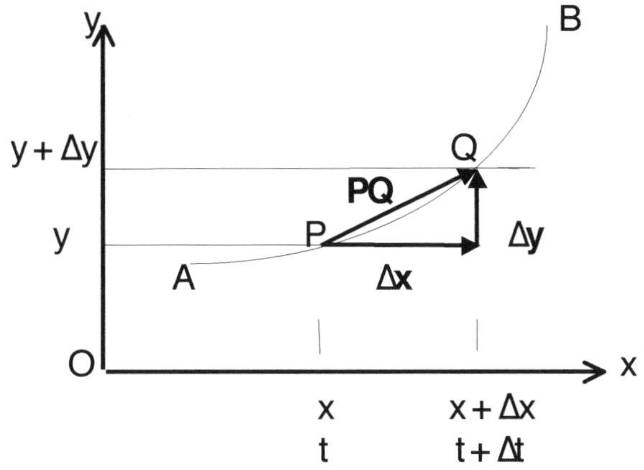

P and Q are the positions of a point moving along the curved path AB. The coordinates are P [x, y] at time t and Q [x + Δx, y + Δy] at time t + Δt. The components of the velocity of the point are

$$\lim(\Delta t \to 0)\ \Delta x/\Delta t = dx/dt = v_x,$$

and

$$\lim(\Delta t \to 0)\ \Delta y/\Delta t = dy/dt = v_y.$$

Δx and Δy are the components of the vector **PQ**. The velocity is therefore

$$\lim(\Delta t \to 0)\ \text{chord}PQ/\Delta t\,.$$

We have

$$\lim(Q \to P)\ \text{chord } PQ/\Delta s = 1,$$

where s is the length of the curve AP and Δs is the length of the arc PQ.

The velocity can be written

$$\lim(\Delta t \rightarrow 0)\ (\text{chord}PQ/\Delta s)(\Delta s/\Delta t) = ds/dt. \qquad (7.1)$$

The direction of the instantaneous velocity at P is along the tangent to the path at P.

The x- and y-components of the acceleration of P are

$$\lim(\Delta t \rightarrow 0)\ \Delta v_x/\Delta t = dv_x/dt = d^2x/dt^2,$$

and

$$\lim(\Delta t \rightarrow 0)\ \Delta v_y/\Delta t = dv_y/dt = d^2y/dt^2.$$

The resultant acceleration is not directed along the tangent at P.

Consider the motion of P along the curve APQB:

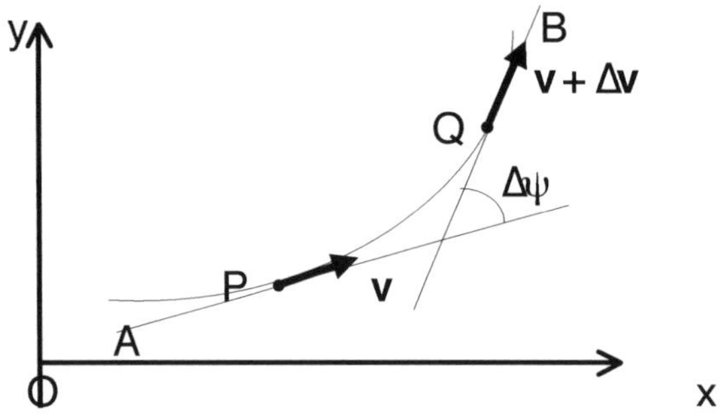

The change $\Delta\mathbf{v}$ in the vector $\mathbf{v}$ is shown in the diagram:

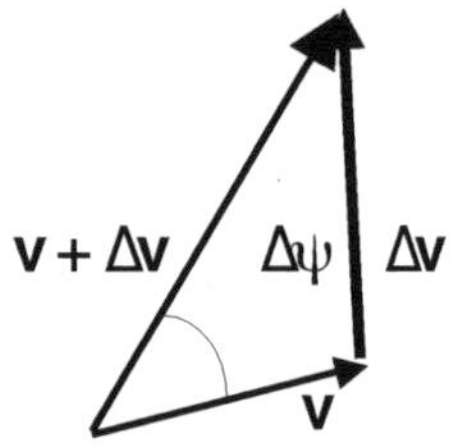

The vector $\Delta\mathbf{v}$ can be written in terms of two components, **a**, perpendicular to the direction of **v,** and **b**, along the direction of $\mathbf{v} + \Delta\mathbf{v}$: The acceleration is

$$\lim(\Delta t \rightarrow 0)\ \Delta\mathbf{v}/\Delta t,$$

The component along **a** is

$$\lim(\Delta t \rightarrow 0)\ \Delta a/\Delta t = \lim(\Delta t \rightarrow 0)\ v\Delta\psi/\Delta t = \lim(\Delta t \rightarrow 0)\ (v\Delta\psi/\Delta s)(\Delta s/\Delta t)$$

$$= v^2(d\psi/ds) = v^2/\rho \qquad (7.2)$$

where

$$\rho = ds/d\psi, \text{ is the radius of curvature at P.} \qquad (7.3)$$

The direction of this component of the acceleration is along the inward normal at P.

If the particle moves in a circle of radius R then its acceleration towards the center is v^2/R, a result first given by Newton.

The component of acceleration along the tangent at P is $dv/dt = v(dv/ds) = d^2s/dt^2$.

7.2 An overview of Newtonian gravitation

Newton considered the fundamental properties of motion, embodied in his three Laws, to be universal in character — the natural laws apply to all motions of all particles throughout all space, at all times. Such considerations form the basis of a Natural Philosophy. In the *Principia*, Newton wrote …"I began to think of gravity as extending to the orb of the Moon…" He reasoned that the Moon, in its steady orbit around the Earth, is always accelerating towards the Earth. He estimated the acceleration as follows:

If the orbit of the Moon is circular (a reasonable assumption), the dynamical problem is

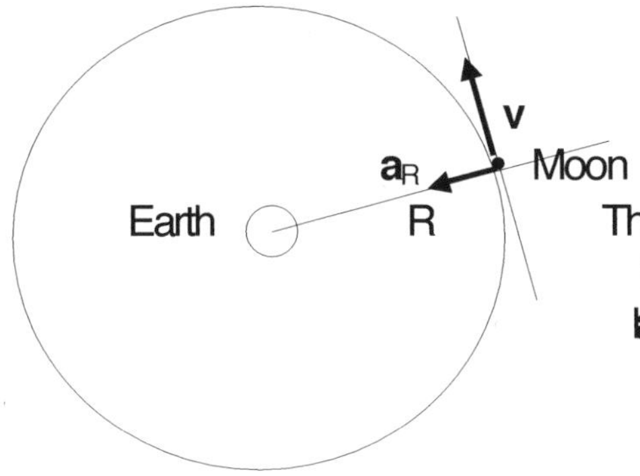

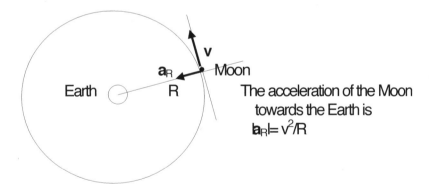

The acceleration of the Moon towards the Earth is $|a_R| = v^2/R$

Newton calculated $v = 2\pi R/T$, where R =240,000 miles, and T = 27.4 days, the period,

so that

$$a_R = 4\pi^2 R/T^2 \approx 0.007 \text{ ft/sec}^2. \qquad (7.4)$$

He knew that all objects, close to the surface of the Earth, accelerate towards the Earth with a value determined by Galileo, namely $g \approx 32$ ft/sec^2. He was therefore faced with the problem of explaining the origin of the very large difference between the value of the acceleration a_R, nearly a quarter of a million miles away from Earth, and the local value, g.

He had previously formulated his 2nd Law that relates *force* to *acceleration*, and therefore he reasoned that the difference between the accelerations, a_R and g, must be associated with a *property of the force* acting between the Earth and the Moon — the force must decrease in some unknown way.

Newton then introduced his conviction that the force of gravity between objects is a *universal* force; each planet in the solar system interacts with the Sun via the same basic force, and therefore undergoes a characteristic acceleration towards the Sun. He concluded that the answer to the problem of the nature of the gravitational force must be contained in the three empirical Laws of Planetary Motion announced by Kepler, a few decades before. The three Laws are

1) The planets describe ellipses about the Sun as focus,
2) The line joining the planet to the Sun sweeps out equal areas in equal intervals of time, and
3) The period of a planet is proportional to the length of the semi-major axis of the orbit, raised to the power of 3/2.

These remarkable Laws were deduced after an exhaustive study of the motion of the planets, made over a period of about 50 years by Tycho Brahe and Kepler.

The 3rd Law was of particular interest to Newton because it relates the square of the period to the cube of the radius for a circular orbit:

$$T^2 \propto R^3 \qquad (7.5)$$

or

$$T^2 = CR^3,$$

where C is a constant. He replaced the specific value of (R/T^2) that occurs in the expression for the acceleration of the Moon towards the Earth with the value obtained from Kepler's 3rd Law and obtained a value for the acceleration a_R:

$$a_R = v^2/R = 4\pi^2 R/T^2 \text{ (Newton)} \tag{7.6}$$

but

$$R/T^2 = 1/CR^2 \text{ (Kepler)} \tag{7.7}$$

therefore

$$a_R = 4\pi^2(R/T^2)$$

$$= (4\pi^2/C)(1/R^2) \text{ (Newton).} \tag{7.8}$$

The acceleration of the Moon towards the Earth varies as the *inverse square* of the distance between them. Newton was now prepared to develop a general theory of gravitation. If the acceleration of a planet towards the Sun depends on the inverse square of their separation, then the force between them can be written, using the 2nd Law of Motion, as follows

$$F = M_{planet}\, a_{planet} = M_{planet}(4\pi^2/C)(1/R^2). \tag{7.9}$$

At this point, Newton introduced the first *symmetry* argument in Physics: if the planet experiences a force from the Sun then the Sun must experience the same force from the planet (the 3rd Law of Motion!). He therefore argued that the expression for the force between the planet and the Sun must contain, explicitly, the masses of the planet *and* the Sun. The gravitational force F_G between them therefore has the form

$$F_G = GM_{Sun}M_{planet}/R^2, \tag{7.10}$$

where G is a constant.

Newton saw no reason to limit this form to the Sun-planet system, and therefore he announced that for *any* two spherical masses, M_1 and M_2, the gravitational force between them is given by

$$F_G = GM_1M_2/R^2, \qquad (7.11)$$

where G is a universal constant of Nature.

All evidence points to the fact that the gravitational force between two masses is always *attractive*.

Returning to the Earth-Moon system, the force on the Moon (mass M_M) in orbit is

$$F_R = GM_EM_M/R^2 = M_Ma_R \qquad (7.12)$$

so that

$a_R = GM_E/R^2$, which is independent of M_M. (The cancellation of the mass M_M in the expressions for F_R involves an important point that is discussed later in the section **8.1**).

At the surface of the Earth, the acceleration, g, of a mass M is essentially constant. It does not depend on the value of the mass, M, thus

$$g = GM_E/R_E^2, \text{ where } R_E \text{ is the radius of the Earth.} \qquad (7.13)$$

(It took Newton many years to *prove* that the entire mass of the Earth, M_E, is equivalent to a point mass, M_E, located at the center of the Earth when calculating the Earth's gravitational interaction with a mass on its surface. This result depends on the exact $1/R^2$-nature of the force).

The ratio of the accelerations, a_R/g, is therefore

$$a_R/g = (GM_E/R^2)/(GM_E/R_E^2) = (R_E/R)^2. \qquad (7.14)$$

Newton knew from observations that the ratio of the radius of the Earth to the radius of the Moon's orbit is about 1/60, and therefore he obtained

$$a_R/g \approx (1/60)^2 = 1/3600.$$

so that

$$a_R = g/3600 = (32/3600)\text{ft/sec}^2 = 0.007...\text{ft/sec}^2.$$

In one of the great understatements of analysis, Newton said, in comparing this result with the value for a_R that he had deduced using $a_R = v^2/R$ ..."that it agreed pretty nearly" ...The discrepancy came largely from the errors in the observed ratio of the radii.

7.3 Gravitation: an example of a central force

Central forces, in which a particle moves under the influence of a force that acts on the particle in such a way that it is always directed towards a single point — the center of force — form an important class of problems . Let the center of force be chosen as the origin of coordinates:

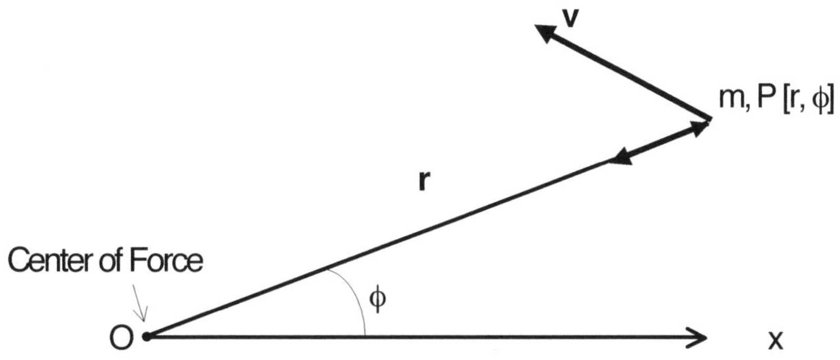

The description of particle motion in terms of polar coordinates (Chapter **2**), is well-suited to the analysis of the central force problem. For general motion, the acceleration of a point P [r, ϕ] moving in the plane has the following components in the r- and "ϕ"- directions

$$\mathbf{a}_r = \mathbf{u}_r(d^2r/dt^2 - r(d\phi/dt)^2), \tag{7.15}$$

and

$$\mathbf{a}_\phi = \mathbf{u}_\phi(r(d^2\phi/dt^2) + 2(dr/dt)(d\phi/dt)), \tag{7.16}$$

where $\mathbf{u}_r$ and $\mathbf{u}_\phi$ are unit vectors in the r- and ϕ-directions.

In the central force problem, the force **F** is always directed towards O, and therefore the component $\mathbf{a}_\phi$, perpendicular to **r**, is always zero:

$$\mathbf{a}_\phi = \mathbf{u}_\phi(r(d^2\phi/dt^2) + 2(dr/dt)(d\phi/dt) = 0, \quad (7.17)$$

and therefore

$$r(d^2\phi/dt^2) + 2(dr/dt)(d\phi/dt) = 0. \quad (7.18)$$

This is the equation of motion of a particle moving under the influence of a central force, centered at O.

If we take the Sun as the (fixed) center of force, the motion of a planet moving about the Sun is given by this equation. The differential equation can be solved by making the substitution

$$\omega = d\phi/dt, \quad (7.19)$$

giving

$$rd\omega/dt + 2\omega(dr/dt) = 0, \quad (7.20)$$

or

$$rd\omega = -2\omega dr.$$

Separating the variables, we obtain

$$d\omega/\omega = -2dr/r.$$

Integrating, gives

$$\log_e\omega = -2\log_e r + C \text{ (constant)},$$

therefore

$$\log_e(\omega r^2) = C.$$

Taking antilogs gives

$$r^2\omega = r^2(d\phi/dt) = e^C = k, \text{ a constant.} \quad (7.21)$$

7.4 Motion under a central force and the conservation of angular momentum

The above solution of the equation of motion of a particle of mass m, moving under the influence of a central force at the origin, O, can be multiplied throughout by the mass m to give

$$mr^2(d\phi/dt) = mk \tag{7.22}$$

or

$$mr(r(d\phi/dt)) = K, \text{ a constant for a given mass,} \tag{7.23}$$

We note that $r(d\phi/dt) = v_\phi$, the component of velocity perpendicular to **r**, therefore

angular momentum of m about O = $r(mv_\phi) = K$, *a constant of the motion for a central force.*

7.5 Kepler's 2nd law explained

The equation $r^2(d\phi/dt)$ = constant, K, can be interpreted in terms of an element of area swept out by the radius vector r, as follows

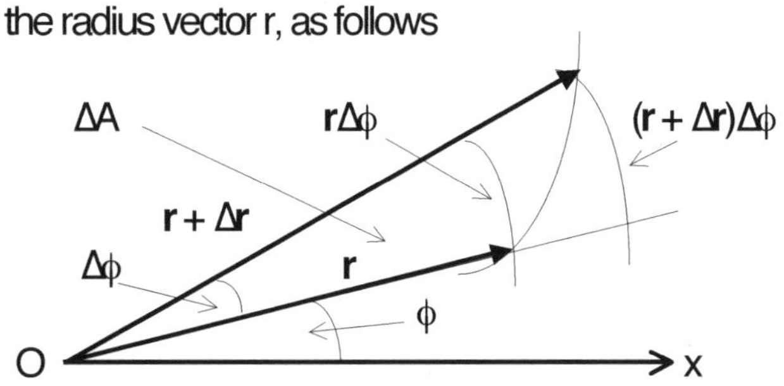

From the diagram, we see that the following inequality holds

$$r^2\Delta\phi/2 < \Delta A < (r + \Delta r)^2\Delta\phi/2$$

or

$$r^2/2 < \Delta A/\Delta\phi < (r + \Delta r)^2/2.$$

When $\Delta\phi \rightarrow 0$, $r + \Delta r \rightarrow r$, so that, in the limit,

$$dA/d\phi = r^2/2.$$

The element of area is therefore

$$dA = r^2 d\phi/2.$$

Twice the time rate of change of this element is therefore

$$2dA/dt = r^2(d\phi/dt). \tag{7.24}$$

We recognize that this expression is equal to k, the constant that occurs in the solution of the differential equation of motion for a central path. The radius vector **r** therefore sweeps out area at a constant rate. This is Kepler's 2nd Law of Planetary Motion; it is seen to be a direct consequence of the fact that the gravitational attraction between the Sun and a planet is a central force problem.

7.6 Central orbits

A central orbit must be a plane curve (there is no force out of the plane), and the moment of the velocity $r^2(d\phi/dt)$, about the center of force, must be a constant of the motion. The moment can be written in three equivalent ways:

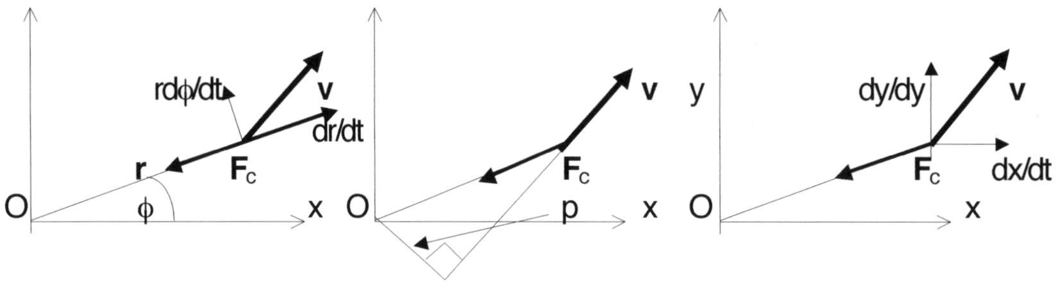

The moment of the velocity about O is then

$$r(r(d\phi/dt) \quad = \quad pv \quad = \quad x(dy/dt) - y(dx/dt)$$

$$= \text{a constant } h\text{, say.} \qquad (7.25)$$

The result $r^2(d\phi/dt)$ = constant for a central force can be derived in the following alternative way:

The time derivative of $r^2(d\phi/dt)$ is

$$(d/dt)(r^2(d\phi/dt)) = r^2(d^2\phi/dt^2) + (d\phi/dt)2r(dr/dt) \qquad (7.26)$$

If this equation is divided throughout by r then

$$(1/r)(d/dt)(r^2(d\phi/dt)) = r(d^2\phi/dt^2) + 2(dr/dt)(d\phi/dt) \qquad (7.27)$$

$$= \text{the transverse acceleration}$$

$$= 0 \text{ for a central force.} \qquad (7.28)$$

Integrating then gives

$$r^2(d\phi/dt) = \text{constant for a central force.} \quad (7.29)$$

7.6.1 The law of force in [p, r] coordinates

There are advantages to be gained in using a new set of coordinates — [p, r] coordinates — in which a point P in the plane is defined in terms of the radial distance r from the origin, and the perpendicular distance p from the origin onto the tangent to the path at P. (See following diagram).

Let a particle of unit mass move along a path under the influence of a central force directed towards a fixed point, O. Let $\mathbf{a}_c$ be the central acceleration of the unit mass at P, let the perpendicular distance from O to the tangent at P be p, and let the instantaneous radius of curvature of the path at the point P be ρ:

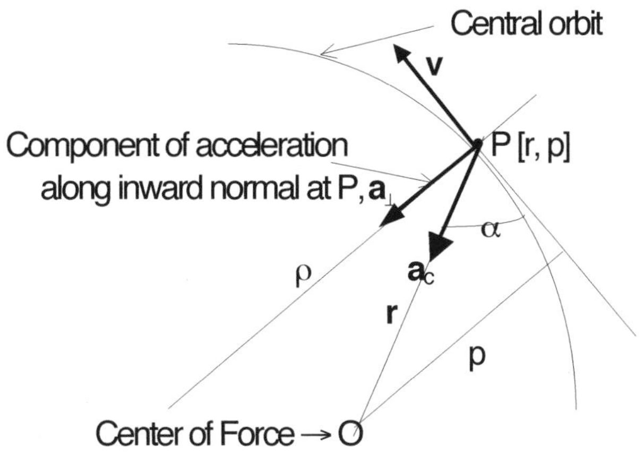

The component of the central acceleration along the inward normal at P is

$$a_\perp = a_c\sin\alpha = v^2/\rho = a_c(p/r). \quad (7.30)$$

The instantaneous radius of curvature is given by

$$\rho = r(dr/dp). \quad (7.31)$$

For *all* central forces,

$$pv = \text{constant} = h, \quad (7.32)$$

therefore

$$a_{\perp} = v^2/\rho = (h^2/p^2)(1/r)(dp/dr) = a_c(p/r), \tag{7.33}$$

so that

$$a_c = (h^2/p^3)(dp/dr). \tag{7.34}$$

This differential equation is the *law of force* per unit mass given the orbit in [p, r] coordinates.

(It is left as a problem to show that given the orbit in polar coordinates, the law of force per unit mass is

$$a_c = h^2u^2\{u + d^2u/d\phi^2\}, \text{ where } u = 1/r \text{).} \tag{7.35}$$

In order to find the law of force per unit mass (acceleration), given the [p, r] equation of the orbit, it is necessary to calculate dp/dr. For example, if the orbit is parabolic, the [p, r] equation can be obtained as follows

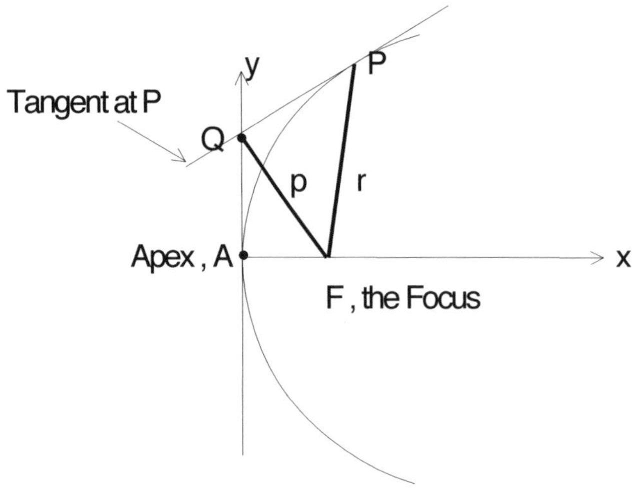

The triangles FAQ and FQP are similar, therefore

$$p/a = r/p, \text{ where AF} = a, \tag{7.36}$$

giving

$$1/p^2 = 1/ar, \text{ the p-r equation of a parabola.} \tag{7.37}$$

Differentiating this equation, we obtain

$$(1/p^3)dp/dr = 1/2ar^2. \tag{7.38}$$

The law of acceleration for the parabolic central orbit is therefore

$$a_c = (h^2/p^3)dp/dr = (h^2/2a)(1/r^2) = \text{constant}/r^2. \quad (7.39)$$

The instantaneous speed of P is given by the equation $v = h/p$; we therefore find

$$v = h/\surd(ar). \quad (7.40)$$

This approach can be taken in discussing central orbits with elliptic and hyperbolic forms.

Consider the ellipse

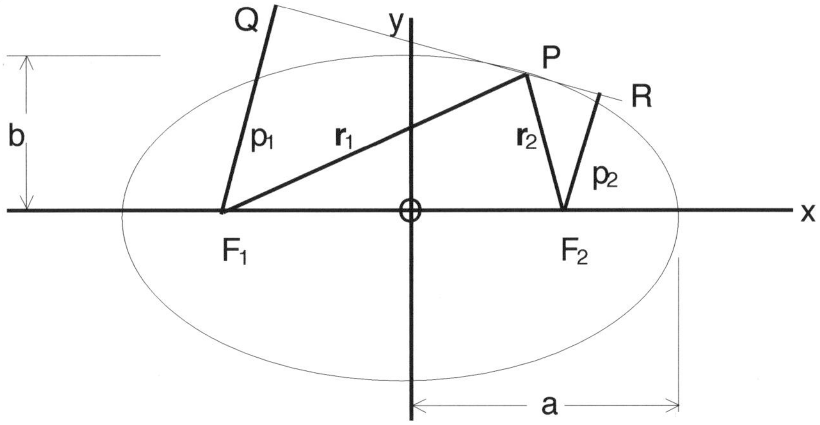

The foci are F_1 and F_2, the semi-major axis is a, the semi-minor axis is b, the radius vectors to the point P [r, ϕ] are $\mathbf{r}_1$ and $\mathbf{r}_2$, and the perpendiculars from F_1 and F_2 onto the tangent at P are p_1 and p_2.

Using standard results from analytic geometry, we have for the ellipse

1) $r_1 + r_2 = 2a$, (7.41 a-c)

2) $p_1p_2 = b^2$, and

3) angle QPF_1 = angle RPF_2.

The triangles F_1QP and F_2RP are similar, and therefore

$$p_1/r_1 = p_2/r_2 \quad (7.42)$$

or

$$(p_1p_2/r_1r_2)^{1/2} = b/\{r_1(2a - r_1)\}^{1/2} = p_1/r_1$$

so that

$$b^2/p_1^2 = 2a/r_1 - 1. \quad (7.43)$$

This is the [p, r] equation of an ellipse.

The [p, r] equation for the hyperbola can be obtained using a similar analysis. The standard results from analytical geometry that apply in this case are

1) $p_1p_2 = b^2$, (7.44 a-c)

2) $r_2 - r_1 = 2a$, and

3) the tangent at P bisects the angle between the focal distances.

($b^2 = a^2(e^2 - 1)$ where e is the eccentricity ($e^2 > 1$), and $2b^2/a$ is the latus rectum).

We therefore obtain

$$b_1^2/p_1^2 = 2a/r_1 + 1. \qquad (7.45)$$

This is the [p, r] equation of an hyperbola.

7.7 Bound and unbound orbits

For a central force, we have the equation for the acceleration in [p, r] form

$$(h^2/p^3)dp/dr = a_c. \qquad (7.46)$$

If the acceleration varies as $1/r^2$, then the form of the orbit is given by separating the variables, and integrating, thus

$$h^2\int dp/p^3 = k\int dr/r^2, \qquad (7.47)$$

so that

$$-h^2/2p^2 = -k/r, \text{ where k is a constant, or}$$

$$h^2/p^2 = 2k/r + C, \text{ where the value of C depends on the form of the orbit.}$$

Comparing this form with the general form of the [p, r] equations of conic sections, we see that the orbit is an ellipse, parabola, or hyperbola depending on the value of C. If

C is negative, the orbit is an ellipse,

C is zero, the orbit is a parabola,

and if

C is positive, the orbit is an hyperbola.

The speed of the particle in a central orbit is given by $v = h/p$. If, therefore, the particle is projected from the origin, O (corresponding to $r = r_0$) with a speed v_0, then

$$h^2/p^2 = v_0^2 = 2k/r_0 + C, \tag{7.48}$$

so that the orbit is

1) an ellipse if $v_0^2 < 2k/r_0$, (7.49 a-c)

2) a parabola if $v_0^2 = 2k/r_0$,

or

3) an hyperbola if $v_0^2 > 2k/r_0$.

The escape velocity, the initial velocity required for the particle to go into an unbound orbit is therefore given by

$v^2_{escape} = 2k/r_0 = 2GM_E/R_E$, for a particle launched from the surface of the Earth. This condition is, in fact, an energy equation

$$\underbrace{(1/2)(m = 1)v^2_{escape}}_{\text{kinetic energy}} = \underbrace{GM_E(m = 1)/R_E}_{\text{potential energy}}. \tag{7.50}$$

7. 8 The concept of the gravitational field

Newton was well-aware of the great difficulties that arise in any theory of the gravitational interaction between two masses not in direct contact with each other. In the *Principia*, he assumes, in the absence of any experimental knowledge of the speed of propagation of the gravitational interaction, that the interaction takes place instantaneously. However, in letters to other luminaries of his day, he postulated an intervening agent between two approaching masses — an agent that requires a *finite time* to react. In the early 17th century, the problem of understanding the interaction between spatially separated objects appeared in a new guise, this time in discussions of the electromagnetic interaction between *charged* objects. Faraday introduced the idea of a *field of force with dynamical properties*. In the Faraday model, an accelerating electric charge acts as the *source* of a dynamical electromagnetic field that travels at a finite speed through space-time, and interacts with a distant charge. Energy and momentum are thereby transferred from one charged object to another distant charged object.

Maxwell developed Faraday's idea into a mathematical theory — the electromagnetic theory of light — in which the speed of propagation of light appears as a fundamental constant of Nature. His theory involves the differential equations of motion of the electric and magnetic field vectors; the equations are not invariant under the Galilean transformation but they are invariant under the Lorentz transformation. (The discovery of the transformation that leaves Maxwell's equations invariant for all inertial observers was made by Lorentz in 1897). We have previously discussed the development of the Special Theory of Relativity by Einstein, *a theory in which there is but one universal constant, c, for the speed of propagation of a dynamical field in a vacuum*. This means that c is not only the speed of light in free space but also the speed of the gravitational field in the void between interacting masses.

We can gain some insight into the dynamical properties associated with the interaction between distant masses by investigating the effect of a finite speed of propagation, c, of the gravitational interaction on Newton's Laws of Motion. Consider a non-orbiting mass M, at a distance R from a mass mass M_S, simply falling from rest with an acceleration **a**(R) towards M_S. According to Newton's Theory of Gravitation, the magnitude of the force on the mass M is

$$|\mathbf{F}(R)| = GM_SM/R^2 = Ma(R), \qquad (7.51)$$

We therefore have

$$a(R) = GM_S/R^2. \qquad (7.52)$$

Let Δt be the time that it takes for the gravitational interaction to travel the distance R at the universal speed c, so that

$$\Delta t = R/c. \qquad (7.53)$$

In the time interval Δt, the mass M moves a distance, ΔR, towards the mass M_S;

$$\begin{aligned} \Delta R &= a\Delta T^2/2 \\ &= (GM_S/R^2)\Delta t^2/2 \\ &= (GM_S/R^2)(R/c)^2/2. \qquad (7.54) \end{aligned}$$

Consider the situation in which the mass M is in a circular orbit of radius R about the mass, M_S. Let **v**(t) be the velocity of the mass M at time t, and **v**(t + Δt) its velocity at t + Δt, where Δt is chosen to be the interaction travel time. Let us consider the motion of M if there were no mass M_S present, and therefore no interaction; the mass M then would continue its motion with constant velocity **v**(t) in a straight line. We are interested in the difference in the positions of M at time t + Δt, with and without the mass M_S in place. We have, to a good approximation:

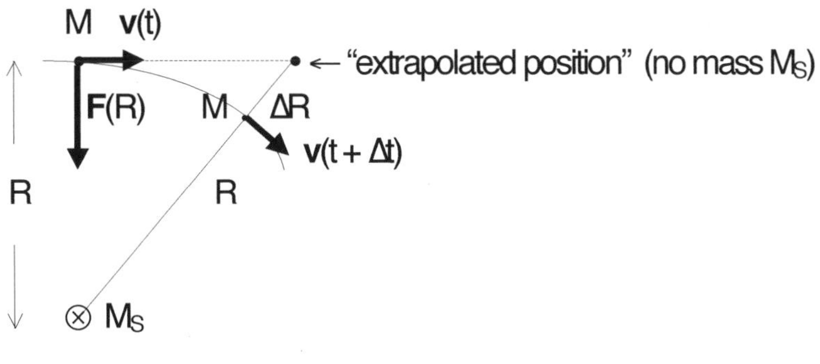

The magnitude of the gravitational force, F_{EX}, at the extrapolated position, with M_S in place, is

$$\begin{aligned} F_{EX} &= GM_SM/(R + \Delta R)^2 \qquad (7.55) \\ &= (GM_SM/R^2)(1 + \Delta R/R)^{-2} \end{aligned}$$

$$\approx (GM_SM/R^2)(1 - 2\Delta R/R), \text{ for } \Delta R \ll R. \tag{7.56}$$

Substituting the value of ΔR obtained above, we find

$$F_{EX} \approx GM_SM/R^2 - (GM_SM/Rc^2)(GM_S/R^2). \tag{7.57}$$

Nerwton's 3rd Law states that

$$\mathbf{F}_{MS,M} = -\mathbf{F}_{M,MS} \tag{7.58}$$

This Law is true, however, for *contact* interactions only. For all interactions that take place between separated objects, there is a mis-match between the action and the reaction. It takes time for one particle to respond to the presence of the other!

In the present example, we obtain a good estimate of the *mismatch* by taking the difference between $F_{EX}(R + \Delta R)$ and $F(R)$, namely

$$F_{EX}(R + \Delta R) - F(R) \approx (GM_SM/Rc^2)(GM_S/R^2). \tag{7.59}$$

On the right-hand side of this equation, we note that the term (GM_S/R^2) has dimensions of "acceleration", and therefore the term (GM_SM/Rc^2) must have dimensions of "mass". We see that this term is an estimate of the "mass" associated with the interaction, itself. The space between the interacting masses must be endowed with this effective mass if Newton's 3rd Law is to include non-contact interactions. The appearance of the term c^2 in the denominator of this effective mass term has a special significance. If we invoke Einstein's famous relation $E = Mc^2$, then $\Delta E = \Delta Mc^2$ so that the effective mass of the gravitational interaction can be written as an effective energy:

$$\Delta E_{GRAV} = GM_SM/R. \tag{7.60}$$

This is the "energy stored in the gravitational field" between the two interacting masses. Note that it has a $1/R$-dependence — the correct form for the *potential energy* associated with a $1/R^2$ gravitational force. We see that the notion of a dynamical field of force is a necessary consequence of the finite propagation time of the interaction.

7.9 The gravitational potential

The concept of a gravitational potential has its origins in the work of Leibniz. The potential energy, V(**x**), asssociated with n interacting particles, of masses m_1, m_2, ...m_n, situated at $\mathbf{x}_1$, $\mathbf{x}_2$, ...$\mathbf{x}_n$, is related to the gravitational force on a mass M at **x**, due to the n particles, by the equation

$$\mathbf{F}(\mathbf{x}) = -\nabla V(\mathbf{x}). \tag{7.61}$$

The exact forms of **F**(**x**) and V(**x**) are

$$\mathbf{F}(\mathbf{x}) = -GM\Sigma_{[i=1,n]}\, m_i(\mathbf{x} - \mathbf{x}_i)/|\mathbf{x} - \mathbf{x}_i|^3, \tag{7.62}$$

and

$$V(\mathbf{x}) = -GM\Sigma_{[i=1,n]}\, m_i/|\mathbf{x} - \mathbf{x}_i|\ .$$

In upper-index notation, the components of the force are

$$F^k(\mathbf{x}) = -\,\partial V/\partial x^k,\ k = 1, 2, 3. \tag{7.63}$$

The gravitational field, **g**(**x**), is the force per unit mass:

$$\mathbf{g}(\mathbf{x}) = \mathbf{F}(\mathbf{x})/M, \tag{7.64}$$

and the gravitational potential is defined as

$$\Phi(\mathbf{x}) = V(\mathbf{x})/M = -\Sigma_{[i=1,n]}\, Gm_i/|\mathbf{x} - \mathbf{x}_i|. \tag{7.65}$$

The sign of the potential is chosen to be negative because the gravitational force is always attractive. (This convention agrees with that used in Electrostatics).

If the mass consists of a continuous distribution that can be described by a mass density $\rho(\mathbf{x})$, then the potential is

$$\Phi(\mathbf{x}) = -\int (G\rho(\mathbf{x}')/|\mathbf{x} - \mathbf{x}'|)\, d^3x'. \tag{7.66}$$

It is left as an exercise to show that this form of Φ means that the potential obeys Poisson's equation

$$\nabla^2\Phi(\mathbf{x}) - 4\pi G\rho(\mathbf{x}) = 0.$$

We should note that the gravitational potential of a mass M has the form

$$V(r) = -GM/r \tag{7.67}$$

only around a mass distribution with spherical symmetry. For an arbitrary mass distribution, the potential can be written as a series of multipoles.

The potential of a circular disc at a point on its axis can be found as follows

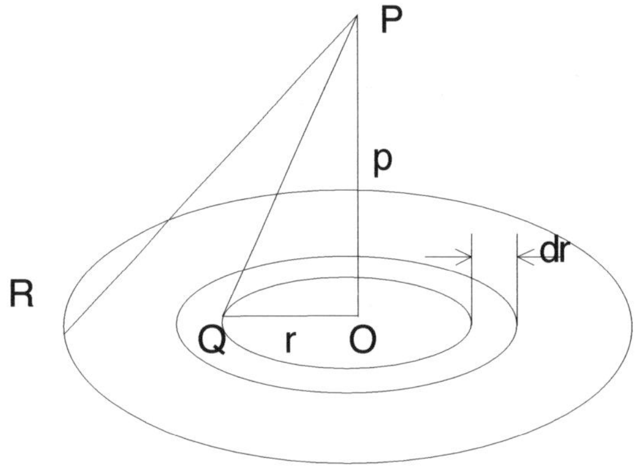

Let the disc be divided into concentric circles. The potential at P, on the axis, due to the elemental ring of radius r and width dr is $2\pi r dr G\sigma/PQ$, where σ is the mass per unit area of the disc. The potential at P of the entire disc is therefore

$$V_P = \int_{[0,a]} 2\pi G\sigma r dr/PQ, \tag{7.68}$$

where a is the radius of the disc. Therefore,

$$V_P = 2\pi G\sigma\int_{[0,a]} r dr/(r^2 + p^2)^{1/2}$$

$$= 2\pi G\sigma[(r^2 + p^2)^{1/2}]_{[0,a]}$$

$$= 2\pi G\sigma(R - p), \tag{7.69}$$

where R is the distance of P from any point on the circumference.

PROBLEMS

7-1 Show that the gravitational potential of a thin spherical shell of radius R and mass M at a point P is

1) GM/d where d is the distance from P to the center of the shell if d >R, and

2) GM/R if P is inside or on the shell.

7-2 If d is the distance from the center of a solid sphere (radius R and density ρ) to a point P inside the sphere, show that the gravitational potential at P is

$$\Phi_P = 2\pi G\rho(R^2 - d^2/3).$$

7-3 Show that the gravitational attraction of a circular disc of radius R and mass per unit area σ, at a point P distant p from the center of the disc, and on the axis, is

$$2\pi G\sigma\{[p/(p^2 + R^2)^{1/2}] - 1\}.$$

7-4 A particle moves in an ellipse about a center of force at a focus. Prove that the instantaneous velocity **v** of the particle at any point in its orbit can be resolved into two components, each of constant magnitude: 1) of magnitude ah/b^2, perpendicular to the radius vector **r** at the point, and 2) of magnitude ahe/b^2 perpendicular to the major axis of the ellipse. Here, a and b are the semi-major and semi-minor axes, e is the eccentricity, and h = pv = constant for a central orbit.

7.5 A particle moves in an orbit under a central acceleration $a = k/r^2$ where k = constant. If the particle is projected with an initial velocity $\mathbf{v}_0$ in a direction at right angles to the radius vecttor **r** when at a distance r_0 from the center of force (the origin), prove

$$(dr/dt)^2 = \{(2k/r_0) - v_o^2(1 + (r_0/r))\}\{(r_0/r) - 1\}.$$

This problem involves the energy and momentum equations in r,ϕ coordinates.

7-6 A particle moves in a cardioidal orbit, $r = a(1 + \cos\phi)$, under a central force

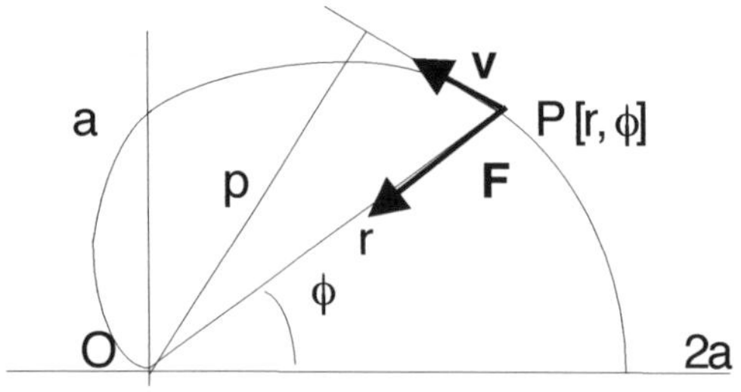

1) show that the p-r equation of the cardioid is $p^2 = r^3/2a$, and

2) show that the central acceleration is $3ah^2/r^4$, where $h = pv$ = constant.

7-7 A planet moves in a circular orbit of radius r about the Sun as focus at the center. If the gravitational "constant" G changes slowly with time — G(t), then show that the angular velocity, ω, of the planet and the radius of the orbit change in time according to the equations

$$(1/\omega)(d\omega/dt) = (2/G)(dG/dt) \text{ and } (1/r)(dr/dt) = (-1/G)(dG/dt).$$

7-8 A particle moves under a central acceleration $a = k(1/r^3)$ where k is a constant. If $k = h^2$, where $h = r^2(d\phi/dt) = pv$, then show that the path is

$1/r = A\phi + B$, a "reciprocal spiral", w

here A and B are constants.

8

EINSTEINIAN GRAVITATION:

AN INTRODUCTION TO GENERAL RELATIVITY

8.1 The principle of equivalence

The term "mass" that appears in Newton's equation for the gravitational force between two interacting masses refers to "gravitational mass" — that property of matter that responds to the gravitational force …Newton's Law should indicate this property of matter:

$F_G = GM^Gm^G/r^2$, where M^G and m^G are the gravitational masses of the interacting objects, separated by a distance r.

The term "mass" that appears in Newton's equation of motion, F = ma, refers to the "inertial mass" — that property of matter that resists changes in its state of motion. Newton's equation of motion should indicate this property of matter:

$F(r) = m^Ia(r)$, where m^I is the inertial mass of the particle moving with an acceleration a(r) in the gravitational field of the mass M^G.

Newton showed by experiment that the inertial mass of an object is equal to its gravitational mass, $m^I = m^G$ to an accuracy of 1 part in 10^3. Recent experiments have shown this equality to be true to an accuracy of 1 part in 10^{12}. Newton therefore took the equations

$$F(r) = GM^Gm^G/r^2 = m^Ia(r), \tag{8.1}$$

and used the condition $m^G = m^I$ to obtain

$$a(r) = GM^G/r^2. \tag{8.2}$$

Galileo had, of course, previously shown that objects made from different materials fall with the same acceleration in the gravitational field at the surface of the Earth, a result that implies $m^G \propto m^I$. This is the Newtonian Principle of Equivalence.

Einstein used this Principle as a basis for a new Theory of Gravitation! He extended the axioms of Special Relativity, that apply to field-free frames, to frames of reference in "free fall". A freely falling frame must be in a state of *unpowered motion in a uniform gravitational field.* The field region must be sufficiently small for there to be no measurable variation in the field throughout the region. If a field gradient does exist in the region then so called "tidal effects" are present, and these can, in principle, be determined (by distorting a liquid drop, for example). The results of all experiments carried out in ideal freely falling frames are therefore fully consistent with Special Relativity. All freely-falling observers measure the speed of light to be c, its constant free-space value. It is not possible to carry out experiments in ideal freely-falling frames that permit a distinction to be made between the acceleration of local, freely-falling objects, and their motion in an equivalent external gravitational field. As an immediate consequence of the extended Principle of Equivalence, Einstein showed that a beam of light would be deflected from its straight path in a dose encounter with a sufficiently massive object. The observers would, themselves, be far removed from the gravitational field of the massive object causing the deflection. Einstein's original calculation of the deflection of light from a distant star, grazing the Sun, as observed here on the Earth, included only those changes in *time intervals* that he had predicted would occur in the near field of the Sun. His result turned out to be in error by exactly a factor of two. He later obtained the "correct" value for the deflection by including in the calculation the changes in *spatial intervals* caused by the gravitational field. A plausible argument is given in the section **8.6** for introducing a non-intuitive concept, the *refractive index of spacetime* due to a gravitational field. This concept is, perhaps, *the* characteristic physical feature of Einstein's revolutionary General Theory of Relativity.

8.2 Time and length changes in a gravitational field

We have previously discussed the changes that occur in the measurement of length and time intervals in different *inertial* frames. These changes have their origin in the invariant speed of light and the necessary synchronization of clocks in a given inertial frame. Einstein showed that measurements of length and time intervals in a given gravitational potential are changed relative to the measurements made in a different gravitational potential. These field-dependent changes are not to be confused with the Special-Relativistic changes discussed in **3.5**. Although an exact treatment of this topic requires the solution of the full Einstein gravitational field equations, we can obtain some of the key results of the theory by making approximations that are valid in the case of our solar system. These approximations are treated in the following sections.

8.3 The Schwarzschild line element

An observer in an ideal freely-falling frame measures an invariant infinitesimal interval of the standard Special Relativistic form

$$ds^2 = (cdt)^2 - (dx^2 + dy^2 + dz^2). \tag{8.3}$$

It is advantageous to transform this form to spherical polar coordinates, using the linear equations

$$x = r\sin\theta\cos\phi,\ y = r\sin\theta\sin\phi,\ \text{and } z = r\cos\theta.$$

We then have

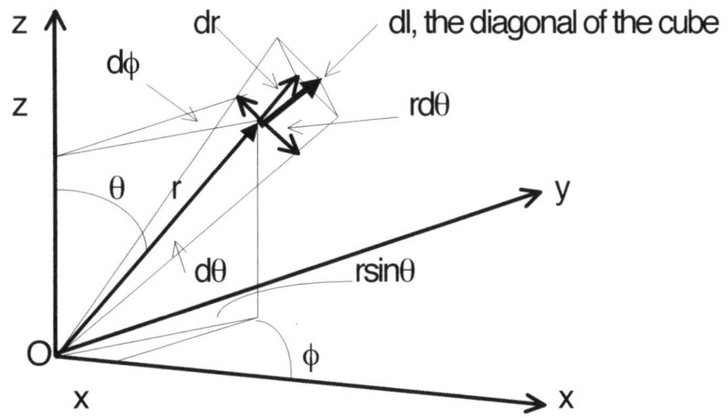

The square of the length of the diagonal of the infinitesimal cube is seen to be

$$dl^2 = dr^2 + (rd\theta)^2 + (r\sin\theta d\phi)^2. \tag{8.4}$$

The invariant interval can therefore be written

$$ds^2 = (cdt)^2 - dr^2 - r^2(d\theta^2 + \sin^2\theta d\phi^2). \tag{8.5}$$

The key question that now faces us is this: how do we introduce gravitation into the problem? We can solve the problem by introducing an energy equation into the argument.

Consider two observers O and O′, passing by one another in a state of *free fall* in a gravitational field due to a mass M, fixed at the origin of coordinates. Both observers measure a standard interval of spacetime, ds according to O, and ds′ according to O′, so that

$$ds^2 = ds'^2 = (cdt')^2 - dr'^2 - r'^2(d\theta'^2 + \sin^2\theta' d\phi'^2) \tag{8.6}$$

The situation is as shown

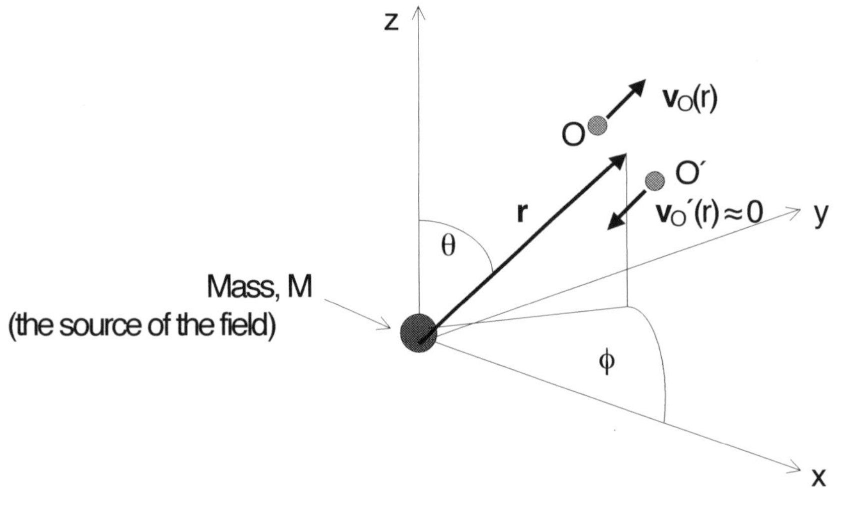

Let the observer O′ just begin free fall towards M at the radial distance r, and let the observer O, close to O′, be *freely falling away* from the mass M. The observer O is in a state of unpowered motion with just the right amount of kinetic energy to "escape to infinity". Since both observers are in states of free fall, we can, according to Einstein, treat them as if they were 'inertial observers". This means that they can relate their local space-time measurements by a Lorentz transformation. In particular, they can relate their measurements of the squared intervals, ds^2 and ds'^2, in the standard way. Since their relative motion is along the radial direction, r, time intervals and radial distances will be measured to be changed:

$$\Delta t = \gamma \Delta t' \text{ and } \gamma \Delta r = \Delta r', \tag{8.7 a, b}$$

where

$$\gamma = 1/\{1 - (v/c)^2\}^{1/2}, \text{ in which } v = v_O(r) \text{ because } v_O'(r) \approx 0.$$

If O has just enough kinetic energy to escape to infinity, then we can equate the kinetic energy to the potential energy, so that

$$v_O^2(r)/2 = 1 \cdot \Phi(r) \text{ if the observer O has unit mass.} \tag{8.8}$$

$\Phi(r)$ is the gravitational potential at r due to the presence of the mass M at the origin.

This procedure enables us to introduce the gravitational potential into the value of γ in the Lorentz transformation. We have $v_O^2 = 2\Phi(r) = v^2$, and therefore

$$\Delta t = \Delta t'/\{1 - 2\Phi(r)/c^2\}^{1/2}, \tag{8.9}$$

and

$$\Delta r = \Delta r'\{1 - 2\Phi(r)/c^2\}^{1/2}. \tag{8.10}$$

Only lengths parallel to **r** change, therefore

$$r^2(d\theta^2 + \sin^2\theta d\phi^2) = r'^2(d\theta'^2 + \sin\theta' d\phi'^2), \tag{8.11}$$

and therefore we obtain

$$ds^2 = ds'^2 = c^2(1 - 2\Phi(r)/c^2)dt^2 - dr^2/(1 - 2\Phi(r)/c^2) - r^2(d\theta^2 + \sin^2\theta d\phi^2). \tag{8.12}$$

If the potential is due to a mass M at the origin then

$$\Phi(r) = GM/r, \quad (r > R, \text{ the radius of the mass, M})$$

therefore,

$$ds^2 = c^2(1 - 2GM/rc^2)dt^2 - (1 - 2GM/rc^2)^{-1}dr^2 - r^2(d\theta^2 + \sin^2\theta d\phi^2). \tag{8.13}$$

This is the famous Schwarzschild line element, originally obtained as an exact solution of the Einstein field equations. The present approach fortuitously gives the exact result!

8.4 The metric in the presence of matter

In the absence of matter, the invariant interval of space-time is

$$ds^2 = \eta_{\mu\nu}dx^\mu dx^\nu \quad (\mu, \nu = 0, 1, 2, 3), \tag{8.14}$$

where

$$\eta_{\mu\nu} = \mathrm{diag}(1, -1, -1, -1) \tag{8.15}$$

is the metric of Special Relativity; it "lowers the indices"

$$dx_\mu = \eta_{\mu\nu}dx^\nu. \tag{8.16}$$

The form of the Schwarzschild line element, ds^2_{sch}, shows that the metric $g_{\mu\nu}$ in the presence of matter differs from $\eta_{\mu\nu}$. We have

$$ds^2_{sch} = g_{\mu\nu}dx^\mu dx^\nu, \tag{8.17}$$

where

$$dx^0 = cdt,\ dx^1 = dr,\ dx^2 = rd\theta,\ \text{and}\ dx^3 = r\sin\theta d\phi,$$

and

$$g_{\mu\nu} = \mathrm{diag}((1 - \chi), -(1 - \chi)^{-1}, -(1 - \chi)^{-1}, -(1 - \chi)^{-1})$$

in which

$$\chi = 2GM/rc^2.$$

The Schwarzschild metric lowers the indices

$$dx_\mu = g_{\mu\nu}dx^\nu, \tag{8.18}$$

so that

$$ds^2_{sch} = dx^{\mu}dx_{\mu}. \tag{8.19}$$

8.5 The weak field approximation

If $\chi = 2GM/rc^2 << 1$, the coefficient, $(1-\chi)^{-1}$, of dr^2 in the Schwarzschild line element can be replaced by the leading term of its binomial expansion, $(1+\chi \ldots)$ to give the "weak field" line element:

$$ds^2_W = (1-\chi)(cdt)^2 - (1+\chi)dr^2 - r^2(d\theta^2 + \sin^2\theta d\phi^2). \tag{8.20}$$

At the surface of the Sun, the value of χ is 4.2×8^{-6}, so that the weak field approximation is valid in all gravitational phenomena in our solar system.

Consider a beam of light traveling radially in the weak field of a mass M, then

$$ds^2_W = 0 \text{ (a light-like interval) , and } d\theta^2 + \sin^2\theta d\phi^2 = 0, \tag{8.21}$$

giving

$$0 = (1-\chi)(cdt)^2 - (1+\chi)dr^2. \tag{8.22}$$

The "velocity" of the light $v_L = dr/dt$, as determined by observers far from the gravitational influence of M, is therefore

$$v_L = c\{(1-\chi)/(1+\chi)\}^{1/2} \neq c \text{ if } \chi \neq 0. \tag{8.23}$$

(Observers in free fall near M always measure the speed of light to be c).

Expanding the term $\{(1-\chi)/(1+\chi)\}^{1/2}$ to first order in χ, we obtain

$$v_L(r)/c \approx (1-\chi/2 \ldots)(1-\chi/2 \ldots)$$

$$= (1-\chi \ldots). \tag{8.24}$$

Therefore

$$v_L(r) \approx c(1-2GM/rc^2 \ldots), \tag{8.25}$$

so that $v_L(r) < c$ in the presence of a mass M according to observers far removed from M.

8.6 The refractive index of space-time in the presence of mass

In Geometrical Optics, the refractive index, n, of a material is defined as

$$n \equiv c/v_{medium} \tag{8.26}$$

where v_{medium} is the speed of light in the medium. We introduce the concept of the *refractive index of space-time*, $n_G(r)$, at a point r in the gravitational field of a mass, M:

$$n_G \equiv c/v_L(r)$$

$$\approx 1/(1-\chi)$$

$$= 1 + \chi \text{ to first-order in } \chi.$$

$$= 1 + 2GM/rc^2. \tag{8.27}$$

The value of n_G increases as r decreases . This effect can be interpreted as an increase in the "density" of space-time as M is approached.

8.7 The deflection of light grazing the sun

As a plane wave of light approaches a spherical mass, those parts of the wave front nearest the mass are slowed down more than those parts farthest from the mass. The speed of the wave front is no longer constant along its surface, and therefore the normal to the surface must be deflected:

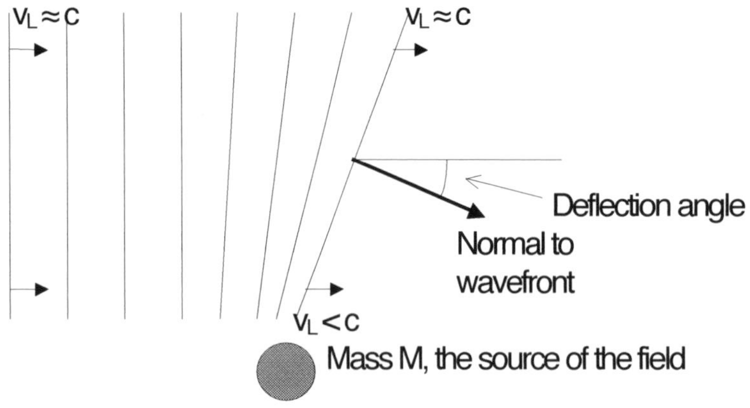

The deflection of a plane wave of light by a spherical mass M as it travels through space-time can be calculated in the weak field approximation. We choose coordinates as shown

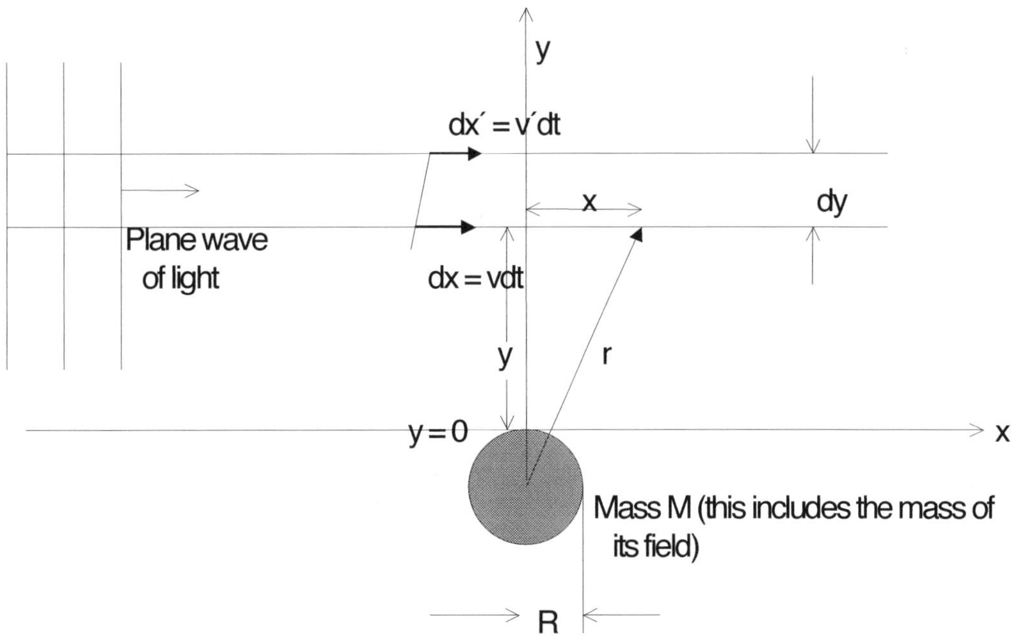

We have shown that the speed of light (moving radially) in a gravitational field, measured by an observer far from the source of the field, depends on the distance, r, from the source

$$v(r) = c(1 - 2GM/rc^2) \qquad (8.28)$$

where c is the invariant speed of light as $r \to \infty$.

We wish to compare dx with dx´, the distances travelled in the x-direction by the wavefront at y and y + dy, in the interval dt.

We have

$$r^2 = (y + R)^2 + x^2 \qquad (8.29)$$

therefore $v(r) \to v(x, y)$ so that

$$2r(\partial r/\partial y) = 2(y + R),$$

and

$$\partial r/\partial y = (y + R)/r . \tag{8.30}$$

Very close to the surface of the mass M (radius R), the gradient is

$$\partial r/\partial y|_{y\to 0} \to R/r. \tag{8.31}$$

Now,

$$\partial v(r)/\partial y = (\partial/\partial r)(c(1 - 2GM/rc^2))(\partial r/\partial y)$$

$$= (2GM/r^2c)(\partial r/\partial y). \tag{8.32}$$

We therefore obtain

$$\partial v(r)/\partial y|_{y\to 0} = (2GM/r^2c)(R/r) = 2GMR/r^3c. \tag{8.33}$$

Let the speed of the wavefront be v' at $y + dy$ and v at y. The distances moved in the interval dt are therefore

$$dx' = v'dt \text{ and } dx = vdt. \tag{8.34 a,b}$$

The first-order Taylor expansion of v' is

$$v' = v + (\partial v/\partial y)dy,$$

and therefore

$$dx' - dx = (v + (\partial v/\partial y)dy)dt - vdt = (\partial v/\partial y)dydt. \tag{8.35}$$

Let the corresponding angle of deflection of the normal to the wavefront be $d\alpha$, then

$$d\alpha = (dx' - dx)/dy$$

$$= (\partial v/\partial y)dt = (\partial v/\partial y)(dx/v). \tag{8.36}$$

The total deflection of the normal to the plane wavefront is therefore

$$\Delta\alpha = \int_{[-\infty,\infty]}(\partial v/\partial y)(dx/v) \tag{8.37}$$

$$\approx (1/c)\int_{[-\infty,\infty]} (\partial v/\partial y)dx .$$

($v \cong c$ over most of the range of the integral).

The portion of the wavefront that grazes the surface of the mass M ($y \rightarrow 0$) therefore undergoes a total deflection

$$\Delta\alpha \approx (1/c)\int_{[-\infty,\infty]}(2GMR/r^3c)dx \quad (8.38)$$

$$= 2GMR/c^2\int_{[-\infty,\infty]}dx/(R^2 + x^2)^{3/2}$$

$$= 2GMR/c^2[x/(R^2(R^2 + x^2)^{1/2})]_{-\infty}^{\infty}$$

$$= 2(GMR/c^2)(2/R^2).$$

so that

$$\Delta\alpha = 4GM/Rc^2.$$

This is Einstein's famous prediction; putting in the known values for G, M, R, and c, gives

$$\Delta\alpha = 1.75 \text{ arcseconds.} \quad (8.39)$$

Measurements of this very small effect, made during total eclipses of the Sun at various times and places since 1919, are fully consistent with Einstein's prediction.

PROBLEMS

8-1 If a particle A is launched with a velocity $\mathbf{v}_{0A}$ from a point P on the surface of the Earth at the same instant that a particle B is dropped from a point Q, use the Principle of Equivalence to show that if A and B are to collide then $\mathbf{v}_{0A}$ must be directed along PQ.

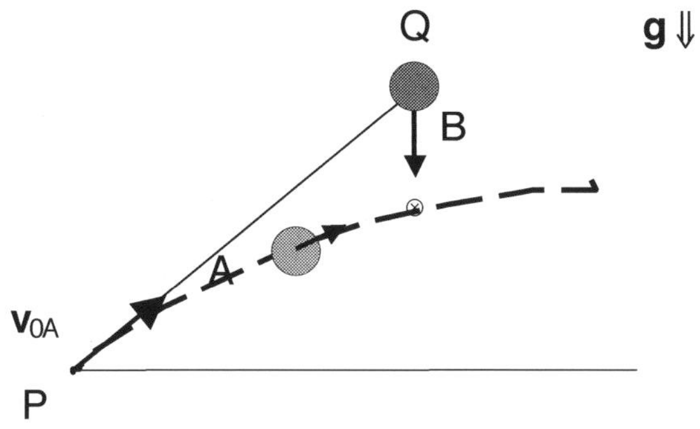

8-2 A satellite is in a circular orbit above the Earth. It carries a clock that is similar to a clock on the Earth. There are two effects that must be taken into account in comparing the rates of the two clocks. 1) the time shift due to their relative speeds (Special Relativity), and 2) the time shift due to their different gravitational potentials (General Relativity). Calculate the SR shift to second-order in (v/c), where v is the orbital speed , and the GR shift to the same order. In calculating the difference in the potentials , integrate from the surface of the Earth to the orbit radius. The two effects differ in sign. Show that the total relative change in the frequency of the satellite clock compared with the Earth clock is

$(\Delta\nu/\nu_E) \approx (gR_E/c^2)\{1 - (3R_E/2r_S)\}$, where r_S is the radius of the satellite orbit (measured from the center of the Earth). We see that if the altitude of the satellite is > $R_E/2$ (~ 3200 km) $\Delta\nu$ is positive since the gravitational effect then predominates, whereas at altitudes less than ~3200 km, the Special Relativity effect predominates. At an altitude ~ 3200 km, the clocks remain in synchronism.

9

AN INTRODUCTION TO THE CALCULUS OF VARIATIONS

9.1 The Euler equation

A frequent problem in Differential Calculus is to find the stationary values (maxima and minima) of a function y(x). The necessary condition for a stationary value at x = a is

$$dy/dx|_{x=a} = 0.$$

For a minimum,

$$d^2y/dx^2|_{x=a} > 0,$$

and for a maximum,

$$d^2y/dx^2|_{x=a} < 0.$$

The Calculus of Variations is concerned with a related problem, namely that of finding a *function* y(x) such that a definite integral taken over a *function of this function* shall be a maximum or a minimum. This is clearly a more complicated problem than that of simply finding the stationary values of a function, y(x).

Explicitly, we wish to find that function y(x) that will cause the definite integral

$$\int_{[x1,\,x2]} F(x, y, dy(x)/dx)dx \qquad (9.1)$$

to have a stationary value.

The integrand F is a function of y(x) as well as of x and dy(x)/dx. The limits x_1 and x_2 are assumed to be fixed , as are the values $y(x_1)$ and $y(x_2)$. The integral has different values along different "paths" that connect (x_1, y_1) and (x_2, y_2). Let a path be Y(x), and let this be one of a set of paths that are adjacent to y(x). We take Y(x) – y(x) to be an infinitesimal for every value of x in the range of integration.

Let the difference be defined as

$$Y(x) - y(x) \equiv \delta y(x) \text{ (a "first-order change")}, \qquad (9.2)$$

and

$$F(x, Y(x), dY(x)/dx) - F(x, y(x), dy(x)/dx) \equiv \delta F. \qquad (9.3)$$

The symbol δ is called a *variation*; it represents the change in the quantity to which it is applied as we go from y(x) to Y(x) *at the same value of x*. Note $\delta x = 0$, and

$$\delta(dy/dx) = dY(x)/dx - dy(x)/dx = (d/dx)(Y(x) - y(x)) = (d/dx)(\delta y(x)).$$

The symbols δ and (d/dx) commute:

$$\delta(d/dx) - (d/dx)\delta = 0. \qquad (9.4)$$

Graphically, we have

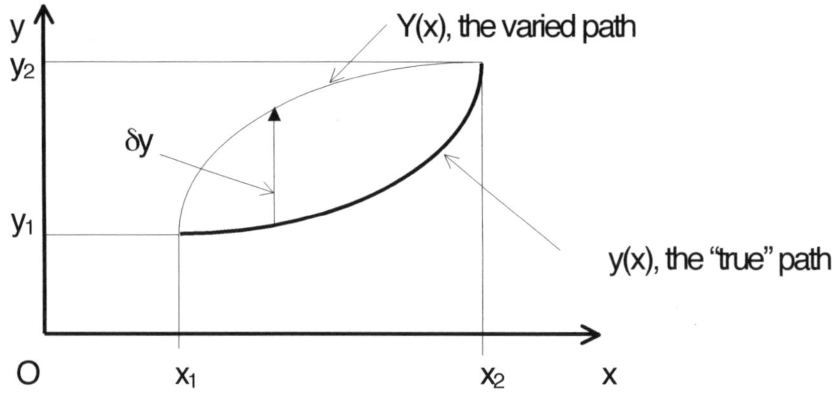

Using the definition of δF, we find

$$\delta F = F(x, \underset{\substack{\uparrow \\ Y(x)}}{\underline{y + \delta y}}, \underset{\substack{\uparrow \\ (d/dx)Y(x)}}{\underline{dy/dx + \delta(dy/dx)}}) - F(x, y, dy/dx) \qquad (9.5)$$

$$= (\partial F/\partial y)\delta y + (\partial F/\partial y')\delta y' \text{ for fixed } x. \text{ (Here, } dy/dx = y'\text{).}$$

The integral

$$\int_{[x1,\, x2]} F(x, y, y')dx, \qquad (9.6)$$

is stationary if its value along the path y is the same as its value along the varied path, $y + \delta y = Y$. We therefore require

$$\int_{[x1,x2]} \delta F(x, y, y')dx = 0. \tag{9.7}$$

This integral can be written

$$\int_{[x1,x2]} \{(\partial F/\partial y)\delta y + (\partial F/\partial y')\delta y'\}dx = 0. \tag{9.8}$$

The second term in this integral can be evaluated by parts, giving

$$[(\partial F/\partial y')\delta y]_{x1}^{x2} - \int_{[x1,x2]} (d/dx)(\partial F/\partial y')\delta y dx. \tag{9.9}$$

But $\delta y_1 = \delta y_2 = 0$ at the end-points x_1 and x_2, therefore the term $[\ldots]_{x1}^{x2} = 0$, so that the stationary condition becomes

$$\int_{[x1,x2]} \{\partial F/\partial y - (d/dx)\partial F/\partial y'\}\delta y dx = 0. \tag{9.10}$$

The infinitesimal quantity δy is positive and arbitrary, therefore, the integrand is zero:

$$\partial F/\partial y - (d/dx)\partial F/\partial y' = 0. \tag{9.11}$$

This is known as Euler's equation.

9.2 The Lagrange equations

Lagrange, one of the greatest mathematicians of the 18th century, developed Euler's equation in order to treat the problem of particle dynamics within the framework of generalized coordinates. He made the transformation

$$F(x, y, dy/dx) \rightarrow L(t, u, du/dt) \tag{9.12}$$

where u is a generalized coordinate and du/dt is a generalized velocity.

The Euler equation then becomes the *Lagrange equation-of-motion*:

$$\partial L/\partial u - (d/dt)(\partial L/\partial \dot{u}) = 0, \text{ where } \dot{u} \text{ is the generalized velocity.} \tag{9.13}$$

The Lagrangian $L(t; u, \dot{u})$ is defined in terms of the kinetic and potential energy of a particle, or system of particles:

$$L \equiv T - V. \tag{9.14}$$

It is instructive to consider the Newtonian problem of the motion of a mass m, moving in the plane, under the influence of an inverse-square-law force of attraction using Lagrange's equations-of-motion. Let the center of force be at the origin of polar coordinates. The kinetic energy of m at $[r, \phi]$ is

$$T = m((dr/dt)^2 + r^2(d\phi/dt)^2)/2, \tag{9.15}$$

and its potential energy is

$$V = -k/r, \text{ where k is a constant.} \tag{9.16}$$

The Lagrangian is therefore

$$L = T - V = m((dr/dt)^2 + r^2(d\phi/dt)^2)/2 + k/r. \tag{9.17}$$

Put r = u, and ϕ = v, the generalized coordinates. We have, for the "u-equation"

$$(d/dt)(\partial L/\partial \dot{u}) = (d/dt)(\partial L/\partial \dot{r}) = (d/dt)(m(dr/dt)) = md^2r/dt^2, \tag{9.18}$$

and

$$\partial L/\partial u = \partial L/\partial r = mr(d\phi/dt)^2 - k/r^2 \tag{9.19}$$

Using Lagrange's equation-of-motion for the u-coordinate, we have

$$m(d^2r/dt^2) - mr(d\phi/dt)^2 + k/r^2 = 0 \tag{9.20}$$

or

$$m(d^2r/dt^2 - r(d\phi/dt)^2) = -k/r^2. \tag{9.21}$$

This is, as it should be, the Newtonian equation

mass $\times$ acceleration in the r-direction = force in the r-direction.

Introducing a second generalized coordinate, we have, for the "v-equation"

$$(d/dt)(\partial L/\partial \dot{v}) = (d/dt)(\partial L/\partial \dot{\phi}) = (d/dt)(mr^2\dot{\phi}) \tag{9.22}$$

$$= m(r^2\ddot{\phi} + \dot{\phi}\,2r\,\dot{r}),$$

and

$$\partial L/\partial v = \partial L/\partial \phi = 0, \quad (9.23)$$

therefore

$$m(r^2\ddot{\phi} + 2r\dot{r}\dot{\phi}) = 0$$

so that

$$(d/dt)(mr^2\dot{\phi}) = 0. \quad (9.24)$$

Integrating , we obtain

$$mr^2\dot{\phi} = \text{constant}, \quad (9.25)$$

showing, again, that the angular momentum is conserved.

The advantages of using the Lagrangian method to solve dynamical problems stem from the fact that L is a *scalar* function of *generalized coordinates*.

9.3 The Hamilton equations

The Lagrangian L is a function of the generalized coordinates and velocities, and the time:

$$L = L(u, v, \ldots; \dot{u}, \dot{v}, \ldots; t). \quad (9.26)$$

If the discussion is limited to two coordinates, u and v, the total differential of L is

$$dL = (\partial L/\partial u)du + (\partial L/\partial v)dv + (\partial L/\partial \dot{u})d\dot{u} + (\partial L/\partial \dot{v})d\dot{v} + (\partial L/\partial t)dt.$$

Consider the simplest case of a mass m moving along the x-axis in a potential, so that $u = x$ and $\dot{u} = \dot{x} = v_x$, then

$$L = T - V = mv_x^2/2 - V \quad (9.27)$$

and

$$\partial L/\partial v_x = mv_x = p_x, \text{ the linear momentum.} \quad (9.28)$$

In general, it is found that terms of the form $\partial L/\partial \dot{u}$ and $\partial L/\partial \dot{v}$ are "momentum" terms;

they are called generalized momenta, and are written

$$\partial L/\partial \dot{u} = p_u,\ \partial L/\partial \dot{v} = p_v,\ \ldots \text{ etc.} \qquad (9.29)$$

Such forms are not limited to "linear" momenta.

The Lagrange equation

$$(d/dt)(\partial L/\partial \dot{u}) - \partial L/\partial u = 0 \qquad (9.30)$$

can be transformed, therefore, into an equation that involves the generalized momenta:

$$(d/dt)(p_u) - \partial L/\partial u = 0, \text{ or}$$

$$\partial L/\partial u = \dot{p}_u. \qquad (9.31)$$

The total differential of L is therefore

$$dL = \dot{p}_u du + \dot{p}_v dv + p_u d\dot{u} + p_v d\dot{v} + (\partial L/\partial t)dt. \qquad (9.32)$$

We now introduce an important function, the Hamiltonian function, H, defined by

$$H \equiv p_u\dot{u} + p_v\dot{v} - L, \qquad (9.33)$$

therefore

$$dH = \{p_u d\dot{u} + \dot{u}dp_u + p_v d\dot{v} + \dot{v}dp_v\} - dL\,. \qquad (9.34)$$

It is not by chance that H is defined in the way given above. The definition permits the cancellation of the terms in dH that involve $d\dot{u}$ and $d\dot{v}$, so that dH depends only on du, dv, dp_u, and dp_v (and perhaps, t). We can therefore write

$$H = f(u, v, p_u, p_v; t) \text{ (limiting the discussion to the two coordinates u and v).} \qquad (9.35)$$

The total differential of H is therefore

$$dH = (\partial H/\partial u)du + (\partial H/\partial v)dv + (\partial H/\partial p_u)dp_u + (\partial H/\partial p_v)dp_v + (\partial H/\partial t)dt. \qquad (9.36)$$

Comparing the two equations for dH, we obtain *Hamilton's equations-of-motion*:

$$\partial H/\partial u = -\dot{p}_u,\ \ \partial H/\partial v = -\dot{p}_v, \qquad (9.37)$$

$$\partial H/\partial p_u = \dot{u},\ \ \partial H/\partial p_v = \dot{v}, \tag{9.38}$$

and

$$\partial H/\partial t = -\partial L/\partial t. \tag{9.39}$$

We see that

$$H = p_u\dot{u} + p_v\dot{v} - (T - V). \tag{9.40}$$

If we consider a mass m moving in the (x, y)-plane then

$$H = (mv_x)v_x + (mv_y)v_y - T + V \tag{9.41}$$

$$= 2(mv_x^2/2 + mv_y^2/2) - T + V$$

$$= T + V, \text{ the total energy.} \tag{9.42}$$

In advanced treatments of Analytical Dynamics, this form of the Hamiltonian is shown to have general validity.

PROBLEMS

9-1 Studies of geodesics — the shortest distance between two points on a surface — form a natural part of the Calculus of Variations. Show that the straight line between two points in a plane is the shortest distance between them.

9-2 The surface generated by revolving the y-coordinate about the x-axis has an area $2\pi\int y ds$ where $ds = \{dx^2 + dy^2\}^{1/2}$ Use Euler's variational method to show that the surface of revolution is a minimum if

$$(dy/dx) = \{(y^2/a^2) - 1\}^{1/2} \text{ where } a = \text{constant.}$$

Hence show that the equation of the minimum surface is

$$y = a\cosh\{(x/a) + b\} \text{ where } b = \text{constant, and } y \neq 0.$$

9-3 The Principle of Least Time pre-dates the Calculus of Variations. The propagation of a ray of light in adjoining media that have different indices of refraction is found to be governed by this principle. A ray of light moves at constant speed v_1 in a medium (1) from a point A to a point B_0 on the x-axis. At B_0, its speed changes to a new constant value v_2 on entering medium (2). The ray continues until it reaches a point C in (2). If the true path $A \to B_0 \to C$ is such that the total travel time of the light in going from A to C is a minimum, show that

$$(v_1/v_2) = x_0\{[y_C^2 + (d - x_0)^2]/[y_A^2 + x_0^2]\}^{1/2}/(d - x_0), \text{ (Snell's law)}$$

where the symbols are defined in the following diagram:

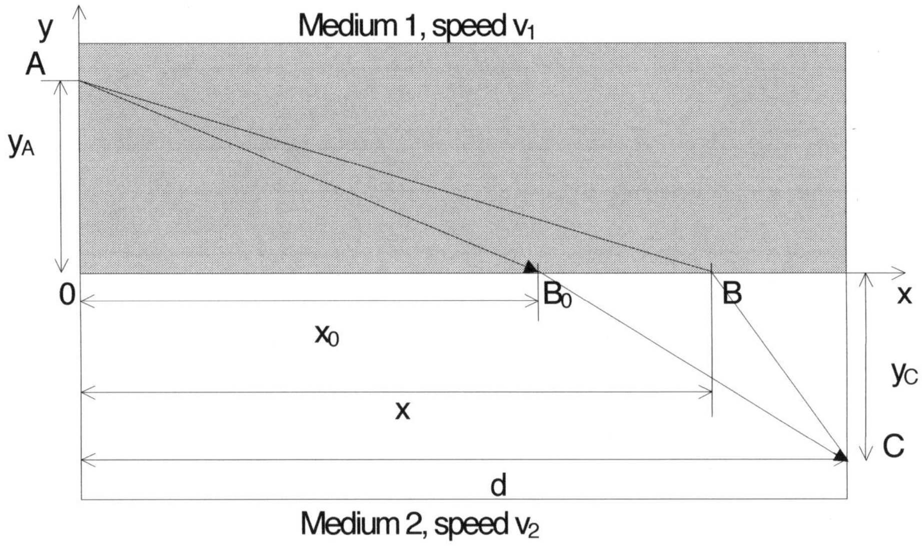

The path $A \to B \to C$ is an arbitrarily varied path.

9-4 Hamilton's Principle states that when a system is moving under conservative forces the time integral of the Lagrange function is stationary. (It is possible to show that this Principle holds for non-conservative forces). Apply Hamilton's Principle to the case of a projectile of mass m moving in a constant gravitational field, in the plane. Let the projectile be launched from the origin of Cartesian coordinates at time $t = 0$. The Lagrangian is

$$L = m((dx/dt)^2 + (dy/dt)^2)/2 - mgy$$

Calculate $\delta\int_{[0,t1]} L dt$, and obtain Newton's equations of motion

$$d^2y/dt^2 + g = 0 \text{ and } d^2x/dt^2 = 0.$$

9-5 Reconsider the example discussed in section **9.2** from the point of view of the Hamiltonian of the system. Obtain $H(r, \phi, p_r, p_\phi)$, and solve Hamilton's equations of motion to obtain the results given in Eqs.9.21 and 9.25.

10

CONSERVATION LAWS, AGAIN

10.1 The conservation of mechanical energy

If the Hamiltonian of a system does not depend explicitly on the time, we have

$$H = H(u, v, \ldots ; p_u, p_v, \ldots). \tag{10.1}$$

In this case, the total differential dH is (for two generalized coordinates, u and v)

$$dH = (\partial H/\partial u)du + (\partial H/\partial v)dv + (\partial H/\partial p_u)dp_u + (\partial H/\partial p_v)dp_v. \tag{10.2}$$

If the positions and the momenta of the particles in the system change with time under their mutual interactions, then H also changes with time, so that

$$dH/dt = (\partial H/\partial u)du/dt + (\partial H/\partial v)dv/dt + (\partial H/\partial p_u)dp_u/dt + (\partial H/\partial p_v)dp_v/dt$$

$$= (-\dot{p}_u\dot{u}) + (-\dot{p}_v\dot{v}) + (\dot{u}\dot{p}_u) + (\dot{v}\dot{p}_v) \tag{10.3}$$

$$= 0, \text{ using Hamilton's equations-of-motion.} \tag{10.4}$$

Integration then gives

$$H = \text{constant.} \tag{10.5}$$

In any system moving under the influence of conservative forces, a potential V exists. In such systems, the total mechanical energy is H = T + V, and we see that it is a constant of the motion.

10.2 The conservation of linear and angular momentum

If the Hamiltonian H does not depend explicitly on a given generalized coordinate then the generalized momentum associated with that coordinate is conserved. For example, if H contains no explicit reference to an *angular coordinate* then the *angular momentum associated with that angle is conserved.* Formally, we have

$$dp_j/dt = -\partial H/\partial q_j, \text{ where } p_j \text{ and } q_j \text{ are the generalized momenta and coordinates.} \tag{10.6}$$

Let an infinitesimal change in the jth-coordinate q_j be made, so that

$$q_j \rightarrow q_j + \delta q_j, \tag{10.7}$$

then we have

$$\delta H = (\partial H/\partial q_j)\delta q_j. \tag{10.8}$$

If the Hamiltonian is *invariant* under the infinitesimal displacement δq_j, then the generalized momentum p_j is a constant of the motion. *The conservation of linear momentum is therefore a consequence of the homogeneity of space, and the conservation of angular momentum is a consequence of the isotropy of space.*

The observed conservation laws therefore imply that the choice of a point in space for the origin of coordinates, and the choice of an axis of orientation play no part in the formulation of the physical laws; the Laws of Nature do not depend on an "absolute space".

11

CHAOS

The behavior of many non-linear dynamical systems as a function of time is found to be chaotic. The characteristic feature of chaos is that the system never repeats its past behavior. Chaotic systems nonetheless obey classical laws of motion which means that the equations of motion are deterministic.

Poincaré was the first to study the effects of small changes in the initial conditions on the evolution of chaotic systems that obey non-linear equations of motion. In a chaotic system, the erratic behavior is due to the internal, or intrinsic, dynamics of the system.

Let a dynamical system be described by a set of first-order differential equations:

$$dx_1/dt = f_1(x_1, x_2, x_3, \ldots x_n) \qquad (11.1)$$

$$dx_2/dt = f_2(x_1, x_2, x_3, \ldots x_n)$$

$$\cdot \quad \cdot$$

$$\cdot \quad \cdot$$

$$dx_n/dt = f_n(x_1, x_2, x_3, \ldots x_n)$$

where the functions f_n are functions of n-variables.

The necessary conditions for chaotic motion of the system are

1) the equations of motion must contain a non-linear term that couples several of the variables.

A typical non-linear equation, in which two of the variables are coupled, is therefore

$$dx_1/dt= ax_1 + bx_2 + cx_1x_2 + \ldots rx_n, \ (a, d, c, \ldots r \text{ are constants}) \qquad (11.2)$$

and

2) the number of independent variables, n, must be at least three.

The second condition is discussed later.

The non-linearity often makes the solution of the equations unstable for particular choices of the parameters. Numerical methods of solution must be adopted in all but a few standard cases.

11.1 The general motion of a damped, driven pendulum

The equation of a damped, driven pendulum is

$$ml(d^2\theta/dt^2) + kml(d\theta/dt) + mg\sin\theta = A\cos(\omega_D t) \tag{11.3}$$

or

$$(d^2\theta/dt^2) + k(d\theta/dt) + (g/l)\sin\theta = (A/ml)\cos(\omega_D t), \tag{11.4}$$

where θ is the angular displacement of the pendulum, l is its length, m is its mass, the resistance is proportional to the velocity (constant of proportionality, k), A is the amplitude and ω_D is the angular frequency of the driving force.

Baker and Gollub in *Chaotic Dynamics* (Cambridge, 1990) write this equation in the form

$$(d^2\theta/dt^2) + (1/q)(d\theta/dt) + \sin\theta = C\cos(\omega_D t), \tag{11.5}$$

where q is the damping factor. The low-amplitude natural angular frequency of the pendulum is unity, and time is dimensionless. We can therefore write

the equation in terms of three first-order differential equations

$$d\omega/dt = -(1/q)\omega - \sin\theta + C\cos(\phi) \text{ where } \phi \text{ is the phase,} \tag{11.6}$$

$$d\theta/dt = \omega, \tag{11.7}$$

and

$$d\phi/dt = \omega_D\,. \tag{11.8}$$

The three variables are (ω, θ, ϕ).

The onset of chaotic motion of the pendulum depends on the choice of the parameters q, C, and ω_D.

The *phase space* of the oscillations is three-dimensional:

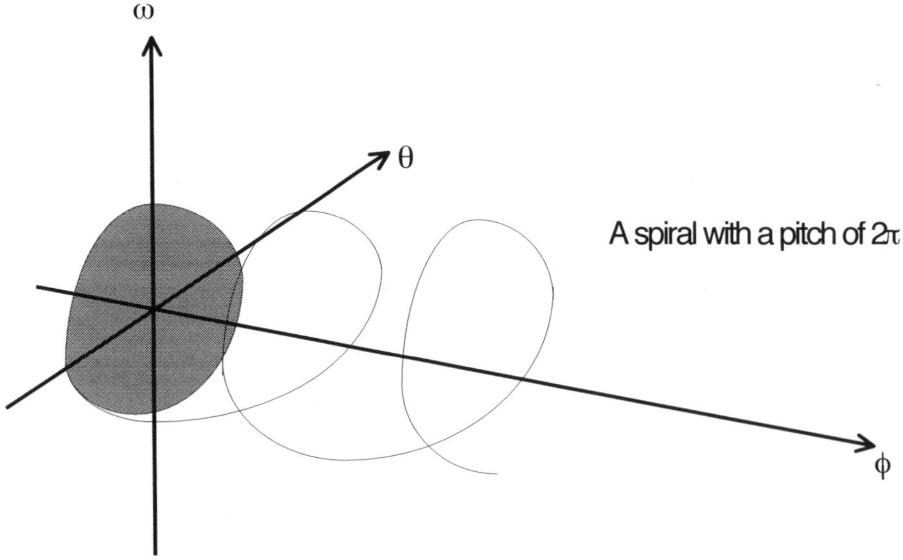

The θ - ω trajectories are projections of the spiral onto the θ - ω plane.

The motion is sensitive to ω_D since the non-linear terms generate many new resonances that occur when $\omega_D/\omega_{natural}$ is a rational number. (Here, $\omega_{natural}$ is the angular frequency of the undamped linear oscillator). For particular values of q and ω_D, the forcing term produces a damped motion that is no longer periodic — the motion becomes chaotic. Periodic motion is characterized by *closed orbits* in the (θ - ω) plane. If the damping is reduced considerably, the motion can become highly chaotic.

The system is sensitive to small changes in the initial conditions. The trajectories in phase space diverge from each other with exponential time-dependence. For chaotic motion, the projection of the trajectory in (θ, ω, φ) - space onto the (θ - ω) plane generates trajectories that intersect. However, in the full 3 - space, a spiraling line along the φ-axis never intersects itself. We therefore see that chaotic motion can exist only when the system has at least a 3 - dimensional phase space. The path then converges towards the *attractor* without self-crossing.

Small changes in the initial conditions of a chaotic system may produce very different trajectories in phase space. These trajectories diverge, and their divergence increases exponentially with time. If the difference between trajectories as a function of time is d(t) then it is found that $\log d(t) \sim \lambda t$ or

$$d(t) \sim e^{\lambda t} \tag{11.9}$$

where $\lambda > 0$ - a positive quantity called the Lyapunov exponent. In a weakly chaotic system $\lambda << 0.1$ whereas, in a strongly chaotic system, $\lambda >> 0.1$.

11.2 The numerical solution of differential equations

A numerical method of solving linear differential equations that is suitable in the present case is known as the Runge-Kutta method. The algorithm for solving two equations that are functions of several variables is:

Let

$$dy/dx = f(x, y, z) \text{ and } dz/dx = g(x, y, z). \tag{11.10}$$

If $y = y_0$ and $z = z_0$ when $x = x_0$ then, for increments h in x_0, k in y_0, and l in z_0

the Runge-Kutta equations are

$k_1 = hf(x_0, y_0, z_0)$ $\quad$ $l_1 = hg(x_0, y_0, z_0)$

$k_2 = hf(x_0 + h/2, y_0 + k_1/2, z_0 + l_1/2)$ $\quad$ $l_2 = hg(x_0 + h/2, y_0 + k_1/2, z_0 + l_1/2)$

$k_3 = hf(x_0 + h/2, y_0 + k_2/2, z_0 + l_2/2)$ $\quad$ $l_3 = hg(x_0 + h/2, y_0 + k_2/2, z_0 + l_2/2)$

$k_4 = hf(x_0 + h, y_0 + k_3, z_0 + l_3)$ $\quad$ $l_4 = hg(x_0 + h, y_0 + k_3, z_0 + l_3)$

$k = (k_1 + 2k_2 + 2k_3 + k_4)/6$

and

$$l = (l_1 + 2l_2 + 2l_3 + l_4)/6. \tag{11.11}$$

The initial values are incremented, and successive values of the x, y, and z are generated by iterations. It is often advantageous to use varying values of h to optimize the procedure.

In the present case,

$$f(x, y, z) \rightarrow f(t, \theta, \omega) \text{ and } g(x, y, z) \rightarrow g(\omega).$$

As a problem, develop an algorithm to solve the non-linear equation 11.5 using the Runge-Kutta method for three equations (11.6, 11.7, and 11.8). Write a program to calculate the necessary iterations. Choose increments in time that are small enough to reveal the details in the θ-ω plane. Examples of non-chaotic and chaotic behavior are shown in the following two diagrams.

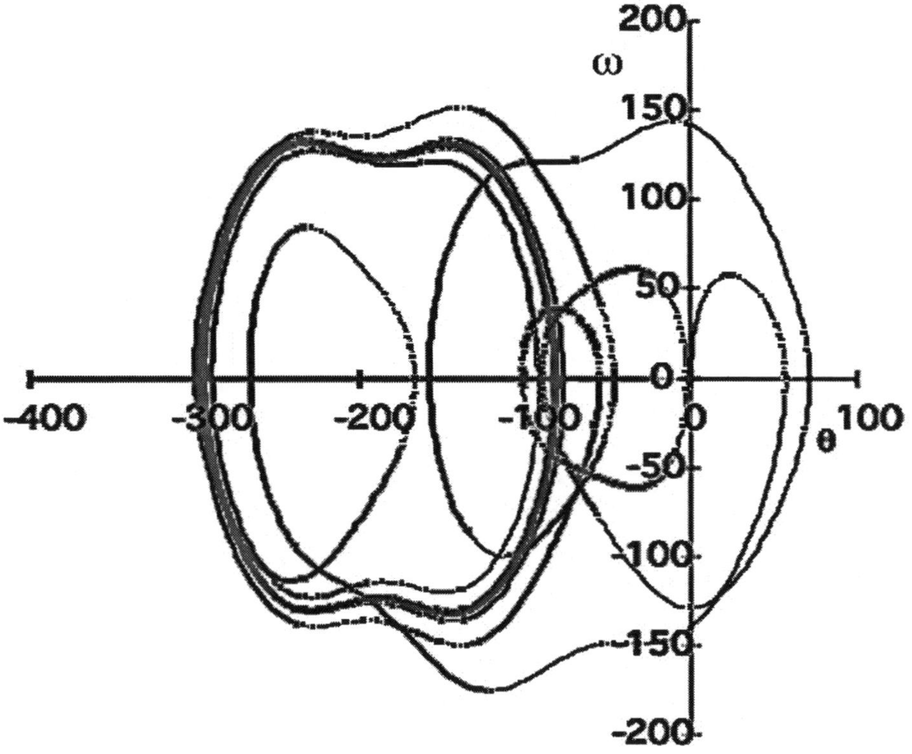

The parameters used to obtain this plot in the θ-ω plane are :

damping factor (1/q) = 1/5,
amplitude (C) = 2,
drive frequency (ω_D) = 0.7, and
time increment, Δt = 0.05.
All the initial values are zero.

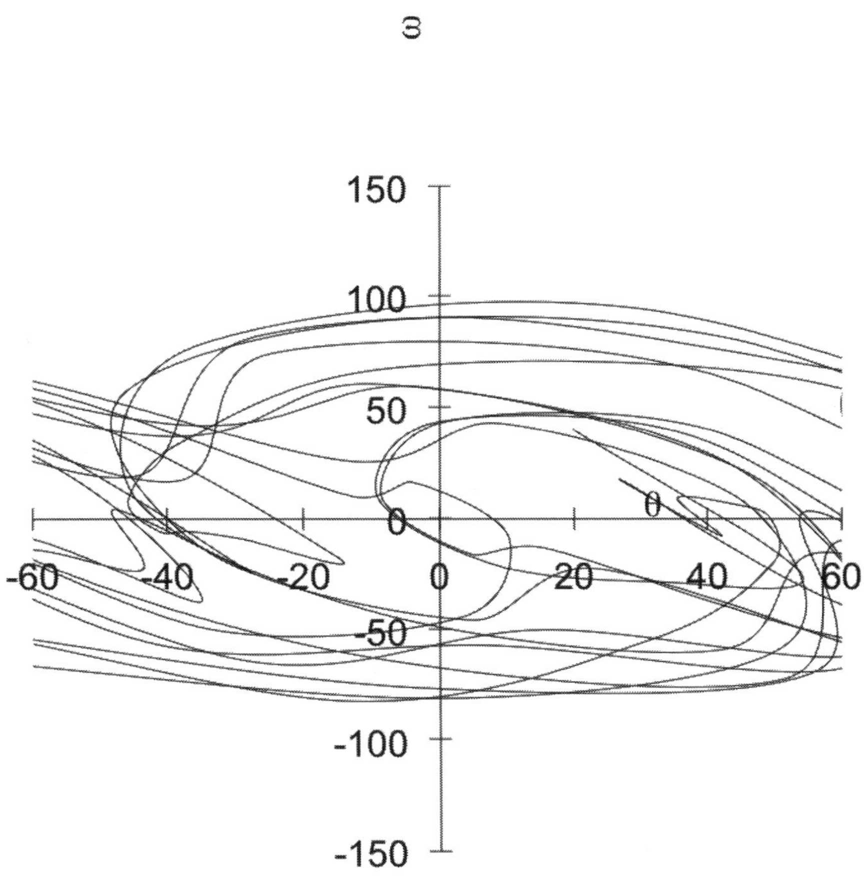

Points in the θ-ω plane for a chaotic system

The parameters used to obtain this plot in the θ-ω plane are:

damping factor (1/q) = 1/2,
amplitude (C) = 1.15,
drive frequency (ω_D) = 0.597, and
time increment, Δt = 1.
The intial value of the time is 100.

12

WAVE MOTION

12.1 The basic form of a wave

Wave motion in a medium is a *collective* phenomenon that involves local interactions among the particles of the medium. Waves are characterized by:

1) a disturbance in space and time.

2) a transfer of energy from one place to another,

and

3) a non-transfer of material of the medium.

(In a water wave, for example, the molecules move perpendicularly to the velocity vector of the wave).

Consider a kink in a rope that propagates with a velocity **V** along the +x-axis, as shown

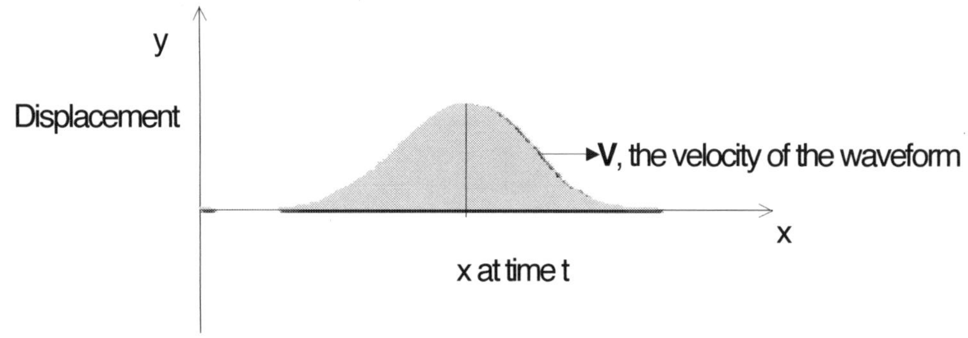

Assume that the shape of the kink does not change in moving a small distance Δx in a short interval of time Δt. The speed of the kink is defined to be $V = \Delta x/\Delta t$. The displacement in the y-direction is a function of x and t,

$$y = f(x, t).$$

We wish to answer the question: what basic principles determine the form of the argument of the function, f ? For water waves, acoustical waves, waves along flexible strings, etc. the wave velocities are much less than c. Since y is a function of x and t, we see that all points on the waveform move in such a way that the Galilean transformation holds for all inertial observers of the waveform. Consider two inertial observers, observer #1 at rest on the x-axis, watching the wave move along the x-axis with constant speed, V, and a second observer #2, moving with the wave. If the observers synchronize their clocks so that $t_1 = t_2 = t_0 = 0$ at $x_1 = x_2 = 0$, then

$$x_2 = x_1 - Vt.$$

We therefore see that the functional form of the wave is determined by the form of the Galilean transformation, so that

$$y(x, t) = f(x - Vt), \quad (12.1)$$

where V is the wave velocity in the particular medium. No other functional form is possible! For example,

$y(x, t) = A\sin k(x - Vt)$ is permitted, whereas

$y(x, t) = A(x^2 + V^2t)$ is not.

If the wave moves to the left (in the –x direction) then

$$y(x, t) = f(x + Vt). \quad (12.2)$$

We shall consider waves that superimpose *linearly*. If, for example, two waves move along a rope in opposite directions, we observe that they "pass through each other".

If the wave is harmonic, the displacement measured as a function of time at the origin, $x = 0$, is also harmonic:

$$y_0(0, t) = A\cos(\omega t)$$

where A is the maximum amplitude, and $\omega = 2\pi\nu$ is the angular frequency.

The general form of y(x, t), consistent with the Galilean transformation, is

$$y(x, t) = A\cos\{k(x - Vt)\}$$

where k is introduced to make the argument dimensionless (k has dimensions of 1/[length]). We then have

$$y_0(0, t) = A\cos(kVt) = A\cos(\omega t).$$

Therefore,

$$\omega = kV, \text{ the angular frequency,} \tag{12.3}$$

or

$$2\pi\nu = kV,$$

so that,

$$k = 2\pi\nu/V = 2\pi/VT \text{ where } T = 1/\nu, \text{ is the period.} \tag{12.4}$$

The general form is then

$$y(x, t) = A\cos\{(2\pi/VT)(x - Vt)\}$$

$$= A\cos\{(2\pi/\lambda)(x - Vt)\}, \text{ where } \lambda = VT \text{ is the wavelength,}$$

$$= A\cos\{(2\pi x/\lambda - 2\pi t/T)\},$$

$$= A\cos(kx - 2\pi t/T), \text{ where } k = 2\pi/\lambda, \text{ the "wavenumber",}$$

$$= A\cos(kx - \omega t),$$

$$= A\cos(\omega t - kx), \text{ because } \cos(-\theta) = \cos(\theta). \tag{12.5}$$

For a wave moving in three dimensions, the displacement at a point x, y, z at time t has the form

$$\psi(x, y, z, t) = A\cos(\omega t - \mathbf{k}\cdot\mathbf{r}), \tag{12.6}$$

where $|\mathbf{k}| = 2\pi/\lambda$ and $\mathbf{r} = [x, y, z]$.

12.2 The general wave equation

An arbitrary waveform in one space dimension can be written as the superposition of two waves, one travelling to the right (+x) and the other to the left (–x) of the origin. The displacement is then

$$y(x, t) = f(x - Vt) + g(x + Vt). \tag{12.7}$$

Put

$$u = f(x - Vt) = f(p), \text{ and } v = g(x + Vt) = g(q),$$

then

$$y = u + v.$$

Now,

$$\partial y/\partial x = \partial u/\partial x + \partial v/\partial x = (du/dp)(\partial p/\partial x) + (dv/dq)(\partial q/\partial x)$$

$$= f'(p)(\partial p/\partial x) + g'(q)(\partial q/\partial x).$$

Also,

$$\partial^2 y/\partial x^2 = (\partial/\partial x)\{(du/dp)(\partial p/\partial x) + (dv/dq)(\partial q/\partial x)\}$$

$$= f'(p)(\partial^2 p/\partial x^2) + f''(p)(\partial p/\partial x)^2 + g'(q)(\partial^2 q/\partial x^2) + g''(q)(\partial q/\partial x)^2.$$

We can obtain the second derivative of y with respect to time using a similar method:

$$\partial^2 y/\partial t^2 = f'(p)(\partial^2 p/\partial t^2) + f''(p)(\partial p/\partial t)^2 + g'(q)(\partial^2 q/\partial t^2) + g''(q)(\partial q/\partial t)^2.$$

Now, $\partial p/\partial x = 1$, $\partial q/\partial x = 1$, $\partial p/\partial t = -V$, and $\partial q/\partial t = V$, and all second derivatives are zero (V is a constant). We therefore obtain

$$\partial^2 y/\partial x^2 = f''(p) + g''(q),$$

and

$$\partial^2 y/\partial t^2 = f''(p)V^2 + g''(q)V^2.$$

Therefore,

$$\partial^2 y/\partial t^2 = V^2(\partial^2 y/\partial x^2).$$

or

$$\partial^2 y/\partial x^2 - (1/V^2)(\partial^2 y/\partial t^2) = 0. \qquad (12.8)$$

This is the wave equation in one-dimensional space. For a wave propagating in three-dimensional space, we have

$$\nabla^2\psi - (1/V^2)(\partial^2\psi/\partial t^2) = 0, \qquad (12.9)$$

the general form of the wave equation, in which $\psi(x, y, z, t)$ is the general amplitude function.

12.3 The Lorentz invariant phase of a wave and the relativistic Doppler shift

A wave propagating through space and time has a "wave function"

$$\psi(x, y, z, t) = A\cos(\omega t - \mathbf{k}\cdot\mathbf{r})$$

where the symbols have the meanings given in **12.2**.

The argument of this function can be written as follows

$$\psi = A\cos\{(\omega/c)(ct) - \mathbf{k}\cdot\mathbf{r}). \qquad (12.10)$$

It was not until deBroglie developed his revolutionary idea of *particle-wave duality* in 1923-24 that the Lorentz invariance of the argument of this function was fully appreciated. We have

$$\psi = A\cos\{[\omega/c, \mathbf{k}]^T[ct, -\mathbf{r}]\}$$

$$= A\cos\{K^{\mu}E_{\mu}\} = A\cos\phi, \text{ where } \phi \text{ is the "phase".} \qquad (12.11)$$

deBroglie recognized that the phase ϕ is a Lorentz invariant formed from the 4-vectors

$$K^{\mu} = [\omega/c, \mathbf{k}], \text{ the "frequency-wavelength" 4-vector,} \qquad (12.12)$$

and

$$E_{\mu} = [ct, -\mathbf{r}], \text{ the covariant "event" 4-vector.}$$

deBroglie's discovery turned out to be of great importance in the development of Quantum Physics. It also provides us with the basic equations for an exact derivation of the *relativistic Doppler shift*. The frequency-wavelength vector is a Lorentz 4-vector, which means that it transforms between inertial observers in the standard way:

$$K^{\mu\prime} = \mathbf{L}K^{\mu}, \qquad (12.13)$$

or

$$\begin{pmatrix} \omega'/c \\ k^{x'} \\ k^{y'} \\ k^{z'} \end{pmatrix} = \begin{pmatrix} \gamma & -\beta\gamma & 0 & 0 \\ -\beta\gamma & \gamma & 0 & 0 \\ 0 & 0 & 1 & 0 \\ 0 & 0 & 0 & 1 \end{pmatrix} \begin{pmatrix} \omega/c \\ k^{x} \\ k^{y} \\ k^{z} \end{pmatrix}$$

The transformation of the first element therefore gives

$$\omega'/c = \gamma(\omega/c) - \beta\gamma k^{x}, \qquad (12.14)$$

so that

$$2\pi\nu' = \gamma 2\pi\nu - \beta\gamma c(2\pi/\lambda)$$

or

$$\nu' = \gamma\nu - V\gamma(\nu/c), \text{ (where } \omega = 2\pi\nu,\ V/c = \beta, \text{ and } c = \nu\lambda)$$

therefore

$$\nu' = \gamma\nu(1 - \beta)$$

or

$$\nu' = (\nu/(1 - \beta^2)^{1/2})(1 - \beta)$$

giving

$$\nu' = \nu\{(1 - \beta)/(1 + \beta)\}^{1/2}. \qquad (12.15)$$

This is the relativistic Doppler shift for the special case of photons – we have Lorentz invariance in action. This result was derived in section **6.2** using the Lorentz invariance of the energy-momentum 4-vector, and the Planck-Einstein result $E = h\nu$ for the relation between the energy E and the frequency ν of a photon. The present derivation of the relativistic Doppler shift is independent of the Planck-Einstein result, and therefore provides an independent verification of their fundamental equation $E = h\nu$ for a photon.

12.4 Plane harmonic waves

The one-dimensional wave equation (12.8) has the solution

$$y(t, x) = A\cos(kx - \omega t),$$

where $\omega = kV$ and A is independent of both x and t.

This form is readily shown to be a solution of (12.8) by direct calculation of its 2nd partial derivatives, and their substitution in the wave equation.

The three-dimensional wave equation (12.9) has the solution

$$\psi(t, x, y, z) = \psi_0\cos\{(k_x x + k_y y + k_z z) - \omega t\},$$

where $\omega = |\mathbf{k}|V$, and $\mathbf{k} = [k_x, k_y, k_z]$, the wave vector.

The solution $\psi(t, x, y, z)$ is called a *plane harmonic wave* because constant values of the argument $(k_x x + k_y y + k_z z) - \omega t$ define a set of planes in space — surfaces of constant phase:

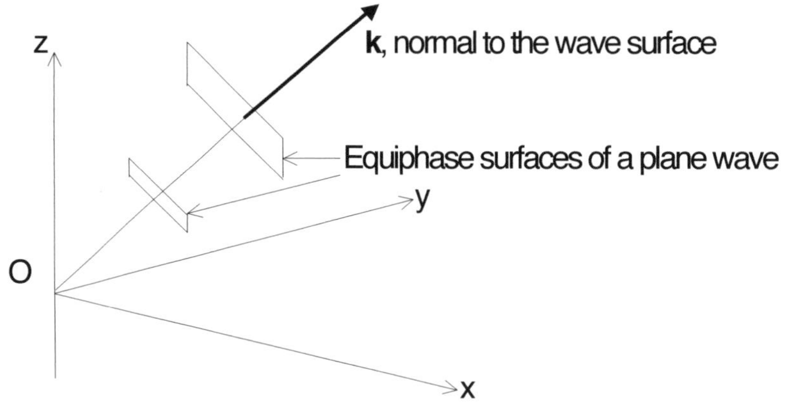

It is often useful to represent a plane harmonic wave as the real part of the remarkable Cotes-Euler equation

$$e^{i\theta} = \cos\theta + i\sin\theta,\ i = \sqrt{-1},$$

so that

$$\psi_0\cos((\mathbf{k}\cdot\mathbf{r}) - \omega t) = \text{R.P. } \psi_0 e^{i(\mathbf{k}\cdot\mathbf{r} - \omega t)}.$$

The complex form is readily shown to be a solution of the three-dimensional wave equation.

12.5 Spherical waves

For given values of the radial coordinate, r, and the time, t, the functions

$\cos(kr - \omega t)$ and $e^{i(kr - \omega t)}$ have constant values on a sphere of radius r. In order for the wave functions to represent expanding spherical waves , we must modify their forms as

follows:

$$(1/r)\cos(kr - \omega t) \text{ and } (1/r)e^{i(kr-\omega t)} \ (\mathbf{k} \text{ along } \mathbf{r}). \quad (12.16)$$

These changes are needed to ensure that the wave functions are solutions of the wave equation. To demonstrate that the spherical wave $(1/r)\cos(kr - \omega t)$ is a solution of (12.9), we must transform the Laplacian operator from Cartesian to polar coordinates,

$$\nabla^2(x, y, z) \rightarrow \nabla^2(r, \theta, \phi).$$

The transformation is

$$\partial^2/\partial x^2 + \partial^2/\partial y^2 + \partial^2/\partial z^2 \rightarrow (1/r^2)[(\partial/\partial r)(r^2(\partial/\partial r)) + (1/\sin\theta)(\partial/\partial\theta)(\sin\theta(\partial/\partial\theta))$$
$$+ (1/\sin^2\theta)(\partial^2/\partial\phi^2)]. \quad (12.17)$$

This transformation is set as a problem.

If there is spherical symmetry, there is no angular-dependence, in which case,

$$\nabla^2(r) = (1/r^2)(\partial/\partial r)(r^2(\partial/\partial r))$$
$$= \partial^2/\partial r^2 + (2/r)(\partial/\partial r). \quad (12.18)$$

We can check that

$$\psi = \psi_0(1/r)\cos(kr - \omega t)$$

is a solution of the radial form of (12.9),

Differentiating twice, we find

$$\partial^2\psi/\partial r^2 = \psi_0[(-k^2/r)\cos u + (2k/r^2)\sin u + (2/r^3)\cos u], \text{ where } u = kr - \omega t,$$

and

$$\partial^2\psi/\partial t^2 = -\psi_0(\omega^2/r)\cos u, \ \omega = kV,$$

from which we obtain

$$(1/V^2)\partial^2\psi/\partial t^2 - [\partial^2\psi/\partial r^2 + (2/r)\partial\psi/\partial r] = 0. \quad (12.19)$$

12.6 The superposition of harmonic waves

Consider two harmonic waves with the same amplitudes, ψ_0, travelling in the same direction, the x-axis. Let their angular frequencies be slightly different — $\omega \pm \delta\omega$ with corresponding wavenumbers $k \pm \delta k$. Their resultant, Ψ, is given by

$$\Psi = \psi_0 e^{i\{(k+\delta k)x-(\omega+\delta\omega)t\}}$$

$$+ \psi_0 e^{i\{(k-\delta k)x-(\omega-\delta\omega)t\}}$$

$$= \psi_0 e^{i(kx-\omega t)}[e^{i(\delta kx-\delta\omega t)} + e^{-i(\delta kx-\delta\omega t)}]$$

$$= \psi_0 e^{i(kx-\omega t)}[2\cos(\delta kx - \delta\omega t)]$$

$$= A\cos\phi, \quad (12.20)$$

where

$$A = 2\psi_0 e^{i(kx-\omega t)}, \text{ the resultant amplitude,}$$

and

$$\phi = \delta kx - \delta\omega t, \text{ the phase of the modulation envelope}.$$

The individual waves travel at a speed

$$\omega/k = v_\phi, \text{ the phase velocity,} \quad (12.21)$$

and the modulation envelope travels at a speed

$$\delta\omega/\delta k = v_G, \text{ the group velocity.} \quad (12.22)$$

In the limit of a very large number of waves, each differing slightly in frequency from that of a neighbor, $dk \to 0$, in which case

$$d\omega/dk = v_G.$$

For electromagnetic waves travelling through a vacuum, $v_G = v_\phi = c$, the speed of light.

We shall not, at this stage, deal with the problem of the superposition of an arbitrary number of harmonic waves.

12.7 Standing waves

The superposition of two waves of the same amplitudes and frequencies but travelling in opposite directions has the form

$$\Psi = \psi_1 + \psi_2 = A\cos(kx - \omega t) + A\cos(kx + \omega t)$$

$$= 2A\cos(kx)\cos(\omega t). \qquad (12.23)$$

This form describes a *standing wave* that pulsates with angular frequency ω, associated with the time-dependent term $\cos\omega t$.

In a traveling wave, the amplitudes of the waves of all particles in the medium are the same and their phases depend on position. In a standing wave, the amplitudes depend on position and the phases are the same.

For standing waves, the amplitudes are a maximum when $kx = 0, \pi, 2\pi, 3\pi, \ldots$

and they are a minimum when $kx = \pi/2, 3\pi/2, 5\pi/2, \ldots$(the nodes).

PROBLEMS

The main treatment of wave motion, including interference and diffraction effects, takes place in the second semester (Part 2) in discussing Electromagnetism and Optics.

12-1 Ripples on the surface of water with wavelengths of about one centimeter are found to have a phase velocity $v_\phi = \sqrt{(\alpha k)}$ where k is the wave number and α is a constant characteristic of water. Show that their group velocity is $v_G = (3/2)v_\phi$.

12-2 Show that

$$y(x, t) = \exp\{x - vt\}$$

represents a traveling wave but not a periodic wave.

12-3 Two plane waves have the same frequency and they oscillate in the z-direction; they have the forms

$$\psi(x, t) = 4\sin\{20t + (\pi x/3) + \pi\}, \text{ and}$$

$$\psi(y, t) = 2\sin\{20t + (\pi y/4) + \pi\}.$$

Show that their superposition at $x = 5$ and $y = 2$ is given by

$$\psi(t) = 2.48\sin\{20t - (\pi/5)\}.$$

12-4 Express the standing wave $y = A\sin(ax)\sin(bt)$, where a and b are constants as a combination of travelling waves.

12-5 Perhaps the most important application of the relativistic Doppler shift has been, and continues to be, the measurement of the velocities of recession of distant galaxies relative to the Earth. The electromagnetic radiation associated with ionized calcium atoms that escape from a galaxy in Hydra has a measured wavlength of 4750×10^{-10}m, and this is to be compared with a wavelength of 3940×10^{-10}m for the same process measured for a stationary source on Earth. Show that the measured wavelengths indicate that the galaxy in Hydra is receding from the Earth with a speed $v = 0.187c$.

13

ORTHOGONAL FUNCTIONS AND FOURIER SERIES

13.1 Definitions

Two n-vectors

$$A_n = [a_1, a_2, ...a_n] \text{ and } B_n = [b_1, b_2, ...b_n]$$

are said to be *orthogonal* if

$$\sum_{[i=1, n]} a_i b_i = 0. \quad (13.1)$$

(Their scalar product is zero).

Two *functions* A(x) and B(x) are *orthogonal* in the range x = a to x = b if

$$\int_{[a, b]} A(x)B(x)dx = 0. \quad (13.2)$$

The limits must be given in order to specify the range in which the functions A(x) and B(x) are defined.

The *set* of real, continuous functions $\{\phi_1(x), \phi_2(x), ...\}$ is orthogonal in [a, b] if

$$\int_{[a, b]} \phi_m(x)\phi_n(x)dx = 0 \text{ for } m \neq n. \quad (13.3)$$

If, in addition,

$$\int_{[a, b]} \phi_n^2(x)dx = 1 \text{ for all } n, \quad (13.4)$$

the set is *normal*, and therefore it is said to be *orthonormal*.

The infinite set

$$\{\cos 0x, \cos 1x, \cos 2x, ... \sin 0x, \sin 1x, \sin 2x, ...\} \quad (13.5)$$

in the range $[-\pi, \pi]$ of x is an example of an orthogonal set. For example,

$$\int_{[-\pi, \pi]} \cos x \cdot \cos 2x dx = 0 \text{ etc.}, \quad (13.6)$$

and

$$\int_{[-\pi, \pi]} \cos^2 x dx \neq 0 = \pi, \text{ etc.}$$

This set, which is orthogonal in any interval of x of length 2π, is of interest in Mathematics because a large class of functions of x can be expressed as linear combinations of the members of the set in the interval 2π. For example we can often write

$$\phi(x) = c_1\phi_1 + c_2\phi_2 + \quad \text{where the c's are constants}$$

$$= a_0\cos0x + a_1\cos1x + a_2\cos2x + ...$$

$$+ b_0\sin0x + b_1\sin1x + b_2\sin2x + ... \qquad (13.7)$$

A large class of *periodic* functions ,of period 2π, can be expressed in this way. When a function can be expressed as a linear combination of the orthogonal set

$$\{1, \cos1x, \cos2x, ...0, \sin1x, \sin2x, ...\},$$

it is said to be *expanded in its Fourier series.*

13.2 Some trigonometric identities and their Fourier series

Some of the familiar trigonometric identities involve Fourier series. For example,

$$\cos2x = 1 - 2\sin^2x \qquad (13.8)$$

can be written

$$\sin^2x = (1/2) - (1/2)\cos2x$$

and this can be written

$$\sin^2x = \{(1/2)\cos0x + 0\cos1x - (1/2)\cos2x + 0\cos3x + ...$$

$$+ 0\{\sin0x + \sin1x + \sin2x + ...\} \qquad (13.9)$$

$$\rightarrow \text{the Fourier series of } \sin^2x.$$

The Fourier series of $\cos^2x$ is

$$\cos^2x = (1/2) + (1/2)\cos2x. \qquad (13.10)$$

More complicated trigonometric identies also can be expanded in their Fourier series. For example, the identity

$$\sin 3x = 3\sin x - 4\sin^3 x$$

can be written

$$\sin^3 x = (3/4)\sin x - (1/4)\sin 3x, \qquad (13.11)$$

and this is the Fourier series of $\sin^3 x$.

The terms in the series represent the "harmonics" of the function $\sin^3 x$.

In a similar fashion, we find that the identity

$$\cos 3x = 4\cos^3 x - 3\cos x$$

can be rearranged to give the Fourier series of $\cos^3 x$

$$\cos^3 x = (3/4) + (1/4)\cos 3x. \qquad (13.12)$$

In general, a combination of deMoivre's theorem and the binomial theorem can be used to write cos(nx) and sin(nx) (for n a positive integer) in terms of powers of sinx and cosx. We have

$$\cos(nx) + i\sin(nx) = (\cos x + i\sin x)^n \quad (i = \sqrt{-1}) \quad \text{(deMoivre)} \qquad (13.13)$$

and

$$(a + b)^n = a^n + na^{n-1}b + (n(n-1)/2!)\, a^{n-2}b^2 \ldots + b^n. \qquad (13.14)$$

For example, if n = 4, we obtain

$$\cos^4 x = (1/8)\cos 4x + (1/2)\cos 2x + (3/8), \qquad (13.15)$$

and

$$\sin^4 x = (1/8)\cos 4x - (1/2)\cos 2x + (3/8). \qquad (13.16)$$

13.3 Determination of the Fourier coefficients of a function

If, in the interval [a, b], the function f(x) can be expanded in terms of the set

$\{\phi_1(x), \phi_2(x), ...\}$, which means that

$$f(x) = \sum_{[i=1,\infty]} c_i\phi_i(x), \tag{13.17}$$

where $\{\phi_1(x), \phi_2(x), ...\}$ is orthogonal in [a, b], then the coefficients can be evaluated as follows:

to determine the kth-coefficient, c_k, multiply f(x) by $\phi_k(x)$, and integrate over the interval [a, b]:

$$\int_{[a,b]} f(x)\phi_k(x)dx = \int_{[a,b]} c_1\phi_1\phi_k dx + \ldots \quad \int_{[a,b]} c_k\phi_k^2 dx + \ldots \tag{13.18}$$

$$= \quad 0 \quad + \quad \ldots \quad + \quad \neq 0 \quad + \ldots \quad 0 \ldots$$

The integrals of the products $\phi_m\phi_n$ in the range $[-\pi, \pi]$ are all zero except for the case that involves ϕ_k^2. We therefore obtain the kth-coefficient

$$c_k = \int_{[a,b]} f(x)\phi_k(x)dx \Big/ \int_{[a,b]} \phi_k^2(x)dx \qquad k = 1, 2, 3, .. \tag{13.19}$$

13.4 The Fourier series of a periodic saw-tooth waveform

In standard works on Fourier analysis it is proved that every periodic continuous function f(x) of period 2π can be expanded in terms of {1, cosx, cos2x, ...0, sinx, sin2x, ...}; this orthogonal set is said to be complete with respect to the set of periodic continuous functions f(x) in [a, b].

Let f(x) be a periodic saw-tooth waveform with an amplitude of ± 1:

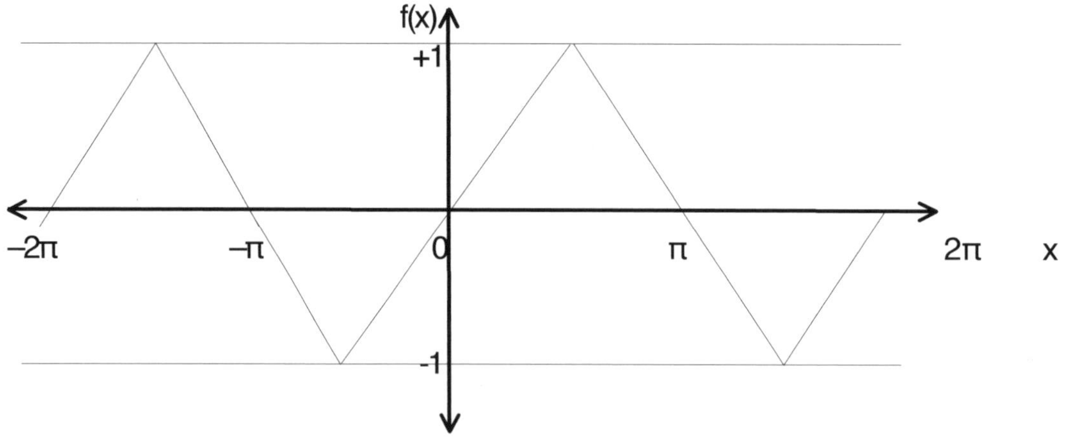

The function has the following forms in the three intervals

$$f(x) = (-2/\pi)(x + \pi) \text{ for } -\pi \le x \le -\pi/2,$$

$= 2x/\pi \qquad \text{for} -\pi/2 \leq x \leq \pi/2,$

and

$= (-2/\pi)(x - \pi) \quad \text{for } \pi/2 \leq x \leq \pi.$

The periodicity means that $f(x + 2\pi) = f(x)$.

The function f(x) can be represented as a linear combination of the series {1, cosx, cos2x, …sinx, sin2x, …}:

$$f(x) = a_0\cos 0x + a_1\cos 1x + a_2\cos 2x + \ldots a_k\cos kx + \ldots$$

$$+ b_0\sin 0x + b_1\sin 1x + b_2\sin 2x + \ldots b_k\sin kx + \ldots \qquad (13.20)$$

The coefficients are given by

$a_k = \int_{[-\pi,\pi]} \cos kx\, f(x)dx / \int_{[-\pi,\pi]} \cos^2 kx dx = 0$, (f(x) is odd, coskx is even, and $[-\pi, \pi]$ is symmetric about 0), (13.21)

and

$b_k = \int_{[-\pi,\pi]} \sin kx\, f(x)dx / \int_{[-\pi,\pi]} \sin^2 kx dx \neq 0,$

$= (1/\pi)\{\int_{[-\pi,-\pi/2]} (-2/\pi)(x + \pi)\sin kx dx + \int_{[-\pi/2,\pi/2]} (2x/\pi)\sin kx dx$

$+ \int_{[\pi/2,\pi]} (-2/\pi)(x - \pi)\sin kx dx\}$

$= \{8/(\pi k)^2\}\sin(k\pi/2). \qquad (13.22)$

The Fourier series of f(x) is therefore

$$f(x) = (8/\pi^2)\{\sin x - (1/3^2)\sin 3x + (1/5^2)\sin 5x - (1/7^2)\sin 7x + \ldots\}.$$

The above procedure can be generalized to include functions that are not periodic. The sum of discrete Fourier components then becomes an integral of the amplitude of the component of angular frequency $\omega = 2\pi\nu$ with respect to ω. This is a subject covered in the more advanced treatments of Physics.

PROBLEMS

13-1 Use deMoivre's theorem and the binomial theorem to obtain the Fourier expansions:

1) $\cos^4x = 3/8 + (1/2)\cos2x + (1/8)\cos4x$,

and

2) $\sin^4x = 3/8 - (1/2)\cos2x + (1/8)\cos4x$.

Plot these components (harmonics) and their sums for $-\pi \leq 0 \leq \pi$.

13-2 Use the method of integration of orthogonal functions to obtain the Fourier series of problem 13-1; you should obtain the same results as above!

13-3 Show that 1) if $f(x) = -f(-x)$, only sine functions occur in the Fourier series for f(x),

and

2) if $f(x) = f(-x)$, only cosine functions occur in the Fourier series for f(x).

13-4 The Fourier series of a function f(t) that is a periodic repetition outside (–T, T), of the shape inside, with period 2π is often written in the form

$f(t) = (a_0/2) + \sum_{[n=1,\infty]}\{a_n\cos(n\pi t/T) + b_n\sin(n\pi t/T)\}$,

where

$a_n = (1/T)\int_{[-T,T]} f(t)\cos(n\pi t/T)dt$,

and

$b_n = (1/T)\int_{[-T,T]} f(t)\sin(n\pi t/T)dt$.

If f(t) is a periodic square-wave:

$f(t) = 3$ for $0 < t < 5\mu s$

$= 0$ for $5 < t < 10\mu s$, with period $2T = 10\mu s$

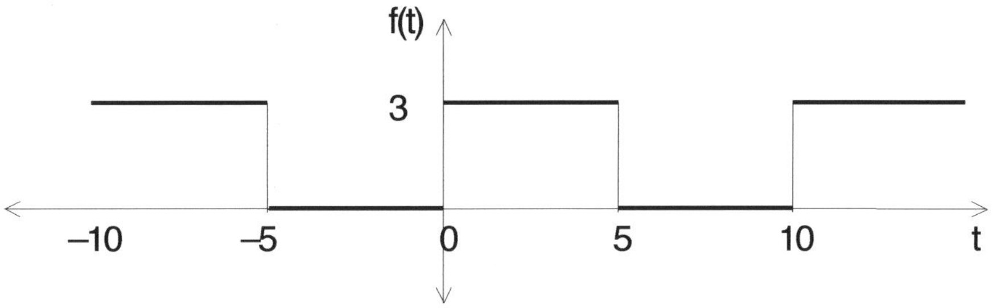

obtain the Fourier series :

$$f(t) = (3/2) + (3/\pi) \sum_{[n=1,\infty]} [(1 - \cos n\pi)/n]\sin(n\pi t/5)).$$

Compute this series for $n = 1$ to 5 and $-5 < t < 5$, and compare the truncated series with the exact waveform.

13-5 It is interesting to note that the series in 13-4 converges to the exact value $f(t) = 3$ at the value $t = 5/2\ \mu s$, so that

$$3 = (3/2) + (3/\pi)\sum_{[n=1,\ \infty]}[(1 - \cos n\pi)/n]\sin(n\pi/2).$$

Use this result to obtain the important Gregory-Leibniz infinite series :

$$(\pi/4) = 1 - (1/3) + (1/5) - (1/7) + \ldots$$

Appendix A

Solving ordinary differential equations

Typical dynamical equations of Physics are

1) Force in the x-direction = mass × acceleration in the x-direction with the mathematical form

$$F_x = ma_x = md^2x/dt^2,$$

and

2) The amplitude y(x, t) of a wave at (x, t), travelling at constant speed V along the x-axis with the mathematical form

$$(1/V^2)\partial^2y/\partial t^2 - \partial^2y/\partial x^2 = 0.$$

Such equations, that involve differential coefficients, are called *differential equations.*

An equation of the form

$$f(x, y(x), dy(x)/dx; a_r) = 0 \tag{A.1}$$

that contains

i) a variable y that depends on a single, independent variable x,

ii) a first derivative dy(x)/dx,

and

iii) constants, a_r,

is called an ordinary (a single independent variable) differential equation of the first order (a first derivative, only).

An equation of the form

$$f(x_1, x_2, ...x_n, y(x_1, x_2, ...x_n), \partial y/\partial x_1, \partial y/\partial x_2, ...\partial y/\partial x_n; \partial^2 y\partial x_1{}^2, \partial^2y/\partial x_2{}^2,$$

$$...\partial^2 y/\partial x_n^{\ 2}; \partial^n y/\partial x_1^{\ n}, \partial^n y/\partial x_2^{\ n}, ...\partial^n y/\partial x_n^{\ n}; a_1, a_2, ...a_r) = 0 \quad (A.2)$$

that contains

i) a variable y that depends on n-independent variables $x_1, x_2, ...x_n$,

ii) the 1st-, 2nd-, ...nth-order partial derivatives:

$\partial y/\partial x_1, ...\partial^2 y/\partial x_1^{\ 2}, ...\partial^n y/\partial x_1^{\ n}, ...,$

and

iii) r constants, $a_1, a_2, ...a_r$,

is called a partial differential equation of the nth-order.

Some of the techniques for solving ordinary linear differential equations are given in this appendix.

An ordinary differential equation is formed from a particular functional relation, $f(x, y; a_1, a_2, ...a_n)$ that involves n arbitrary constants. Successive differentiations of f with respect to x, yield n relationships involving x, y, and the first n derivatives of y with respect to x, and some (or possibly all) of the n constants. There are (n + 1) relationships from which the n constants can be eliminated. The result will involve $d^n y/dx^n$, differential coefficients of lower orders, together with x, and y, and *no* arbitrary constants.

Consider, for example, the standard equation of a parabola:

$y^2 - 4ax. = 0$, where a is a constant.

Differentiating, gives

$2y(dy/dx) - 4a = 0$

so that

$y - 2x(dy/dx) = 0$, a differential equation that does not contain the constant a.

As another example, consider the equation

$f(x, y, a, b, c) = 0 = x^2 + y^2 + ax + by + c = 0.$

Differentiating three times successively, with respect to x, gives

1) $2x + 2y(dy/dx) + a + b(dy/dx) = 0$,

2) $2 + 2\{y(d^2y/dx^2) + (dy/dx)^2\} + b(d^2y/dx^2) = 0$,

and

3) $2\{y(d^3y/dx^3) + (d^2y/dx^2)(dy/dx)\} + 4(dy/dx)(d^2y/dx^2) + b(d^3y/dx^3) = 0$.

Eliminating b from 2) and 3),

$$(d^3y/dx^3)\{1 + (dy/dx)^2\} = (dy/dx)(d^2y/dx^2)^2.$$

The most general solution of an ordinary differential equation of the nth-order contains n arbitrary constants. The solution that contains *all* the arbitrary constants is called the *complete primative*. If a solution is obtained from the complete primative by giving definite values to the constants then the (non-unique) solution is called a *particular integral.*

Equations of the 1st-order and degree.

The equation

$$M(x, y)(dy/dx) + N(x, y) = 0 \tag{A.3}$$

is separable if M/N can be reduced to the form $f_1(y)/f_2(x)$, where f_1 does not involve x, and f_2 does not involve y. Specific cases that are met are:

i) y absent in M and N, so that M and N are functions of x only; Eq. (A.3) then can be written

$$(dy/dx) = -(M/N) = F(x)$$

therefore

$$y = \int F(x)dx + C,\text{ where C is a constant of integration.}$$

ii) x absent in M and N.

Eq. (A.3) then becomes

$$(M/N)(dy/dx) = -1,$$

so that

$$F(y)(dy/dx) = -1, \quad (M/N = F(y))$$

therefore

$$x = -\int F(y)dy + C.$$

iii) x and y present in M and N, but the variables are separable.

Put $M/N = f(y)/g(x)$, then Eq. (A.3) becomes

$$f(y)(dy/dx) + g(x) = 0.$$

Integrating over x,

$$\int f(y)(dy/dx)dx + \int g(x)dx = 0.$$

or

$$\int f(y))dy + \int g(x)dx = 0.$$

For example, consider the differential equation

$$x(dy/dx) + \cot y = 0.$$

This can be written

$$(\sin y/\cos y)(dy/dx) + 1/x = 0.$$

Integrating, and putting the constant of integration $C = \ln D$,

$$\int(\sin y/\cos y)dy + \int(1/x)dx = \ln D,$$

so that

$$-\ln(\cos y) + \ln x = \ln D,$$

or

$$\ln(x/\cos y) = \ln D.$$

The solution is therefore

$$y = \cos^{-1}(x/D).$$

Exact equations

The equation

$y dx + x dy = 0$ is said to be exact because it can be written as

$$d(xy) = 0, \text{ or}$$

$$xy = \text{constant}.$$

Consider the non-exact equation

$$(\tan y)dx + (\tan x)dy = 0.$$

We see that it can be made exact by multiplying throughout by cosxcosy, giving

$$\sin y \cos x dx + \sin x \cos y dy = 0 \text{ (exact)}$$

so that

$$d(\sin y \sin x) = 0,$$

or

$$\sin y \sin x = \text{constant}.$$

The term cosxcosy is called an *integrating factor.*

Homogeneous differential equations.

A homogeneous equation of the nth degree in x and y is such that the powers of x and y in every term of the equation is n. For example, $x^2y + 2xy^2 + 3y^3$ is a homogeneous equation of the third degree. If, in the differential equation $M(dy/dx) + N = 0$ the terms M and N are homogeneous functions of x and y, of the same degree, then we have a homogeneous differential equation of the 1st order and degree. The differential equation then reduces to

$$dy/dx = -(N/M) = F(y/x)$$

To find whether or not a function F(x, y) can be written F(y/x), put

$$y = vx.$$

If the result is F(v) (all x's cancel) then F is homogeneous. For example

$dy/dx = (x^2 + y^2)/2x^2 \rightarrow dy/dx = (1 + v^2)/2 = F(v)$, therefore the equation is homogeneous.

Since $dy/dx \rightarrow F(v)$ by putting $y = vx$ on the right-hand side of the equation, we make the same substitution on the left-hand side to obtain

$$v + x(dv/dx) = (1 + v^2)/2$$

therefore

$$2xdv = (1 + v^2 - 2v)dx.$$

Separating the variables

$2dv/(v - 1)^2 = dx/x.$, and this can be integrated.

Linear Equations

The equation

$$dy/dx + M(x)y = N(x)$$

is said to be linear and of the 1st order. An example of such an equation is

$$dy/dx + (1/x)y = x^2.$$

This equation can be solved by introducing the integrating factor, x, so that

$$x(dy/dx) + y = x^3,$$

therefore

$$(d/dx)(xy) = x^3,$$

giving

$$xy = x^4/4 + \text{constant}.$$

In general, let R be an integrating factor, then

$$R(dy/dx) + RMy = RN,$$

in which case, the left-hand side is the differential coefficient of some product with a first term R(dy/dx). The product must be Ry! Put, therefore

$$R(dy/dx) + RMy = (d/dx)(Ry) = R(dy/dx) + y(dR/dx).$$

Now,

$$RMy = y(dR/dx),$$

which leads to

$$\int M(x)dx = \int dR/R = \ln R,$$

or

$$R = \exp\{\int M(x)dx\}.$$

We therefore have the following procedure: to solve the differential equation

$$(dy/dx) + M(x)y = N(x),$$

multiply each side by the integrating factor $\exp\{\int M(x)dx\}$, and integrate. For example, let

$$(dy/dx) + (1/x)y = x^2,$$

so that

$\int M(x)dx = \int(1/x)dx = \ln x$ and the integrating factor is $\exp\{\ln x\} = x$:. We therefore obtain the equation

$$x(dy/dx) + (1/x)y = x^3,$$

deduced previously on intuitive grounds.

Linear Equations with Constant Coefficients.

Consider the 1st order linear differential equation

$$p_0(dy/dx) + p_1y = 0, \text{ where } p_0, p_1 \text{ are constants.}$$

Writing this as

$$p_0(dy/y) + p_1 dx = 0,$$

we can integrate term-by-term, so that

$$p_0 \ln y + p_1 x = \text{constant},$$

therefore

$$\ln y = (-p_1/p_0)x + \text{constant}$$

$$= (-p_1/p_0)x + \ln A, \text{ say}$$

therefore

$$y = A\exp\{(-p_1/p_0)x\}.$$

Linear differential equations with constant coefficients of the 2nd order occur often in Physics. They are typified by the forms

$$p_0(d^2y/dx^2) + p_1(dy/dx) + p_2y = 0.$$

The solution of an equation of this form is obtained by following the insight gained in solving the 1st order equation!. We try a solution of the type

$$y = A\exp\{mx\},$$

so that the equation is

$$A\exp\{mx\}(p_0m^2 + p_1m + p_2) = 0.$$

If m is a root of

$$p_0m^2 + p_1m + p_2 = 0$$

then $y = A\exp\{mx\}$ is a solution of the original equation for all values of A.

Let the roots be α and β. If $\alpha \neq \beta$ there are two solutions

$$y = A\exp\{\alpha x\} \text{ and } y = B\exp\{\beta x.\}.$$

If we put

$$y = A\exp\{\alpha x\} + B\exp\{\beta x\}$$

in the original equation then

$$A\exp\{\alpha x\}(p_0\alpha^2 + p_1\alpha + p_2) + B\exp\{\beta x\}(p_0\beta^2 + p_1\beta + p_2) = 0,$$

which is true as α and β are the roots of

$$p_0m^2 + p_1m + p_2 = 0,$$ (called the *auxilliary equation*)

The original equation is linear, therefore the sum of the two solutions is, itself, a (third) solution. The third solution contains two arbitrary constants (the order of the equation), and it is therefore the *general solution*.

As an example of the method, consider solving the equation

$$2(d^2y/dx^2) + 5(dy/dx) + 2y = 0.$$

Put $y = A\exp\{mx\}$ as a trial solution, then

$A\exp\{mx\}(2m^2 + 5m + 2) = 0$, so that

$m = -2$ or $-1/2$, therefore the general solution is

$$y = A\exp\{-2x\} + B\exp\{(-1/2)x\}.$$

If the roots of the auxilliary equation are complex, then

$$y = A\exp\{p + iq\}x + B\exp\{p - iq\}x,$$

where the roots are $p \pm iq$ ($p, q \in R$).

In practice, we write

$$y = \exp\{px\}[E\cos qx + F\sin qx]$$

where E and F are arbitrary constants.

For example, consider the solution of the equation

$$d^2y/dx^2 - 6(dy/dx) + 13y = 0,$$

therefore

$$m^2 - 6m + 13 = 0,$$

so that

$$m = 3 \pm i2.$$

We therefore have

$$y = A\exp\{(3 + i2)x\} + B\exp\{3 - i2)x\}$$

$$= \exp\{3x\}(E\cos 2x + F\sin 2x).$$

The general solution of a linear differential equation with constant coefficients is the sum of a *particular integral* and the *complementary function* (obtained by putting zero for the function of x that appears in the original equation).

BIBLIOGRAPHY

Those books that have had an important influence on the subject matter and the style of this book are recognized with the symbol *. I am indebted to the many authors for providing a source of fundamental knowledge that I have attempted to absorb in a process of continuing education over a period of fifty years.

General Physics

*Feynman, R. P., Leighton, R. B., and Sands, M., *The Feynman Lectures on Physics*, 3 vols., Addison-Wesley Publishing Company, Reading, MA (1964).

*Joos, G., *Theoretical Physics*, Dover Publications, Inc., New York, 3rd edn (1986).

Lindsay, R. B., *Concepts and Methods of Theoretical Physics*, Van Nostrand Company, Inc., New York (1952).

Mathematics

Armstrong, M. A., *Groups and Symmetry*, Springer-Verlag, New York (1988).

*Caunt, G. W., *An Introduction to Infinitesimal Calculus*, The Clarendon Press, Oxford (1949).

*Courant R., and John F., *Introduction to Calculus and Analysis*, 2 vols., John Wiley & Sons, New York (1974).

Kline, M., *Mathematical Thought from Ancient to Modern Times*, Oxford University Press, Oxford (1972).

*Margenau, H., and Murphy, G. M., *The Mathematics of Physics and Chemistry*, Van Nostrand Company, Inc., New York, 2nd edn (1956).

Mirsky, L., *An Introduction to Linear Algebra*, Dover Publications, Inc., New York (1982).

*Piaggio, H. T. H., *An Elementary Treatise on Differential Equations*, G. Bell & Sons, Ltd., London (1952).

Samelson, H., *An Introduction to Linear Algebra*, John Wiley & Sons, New York (1974).

Stephenson, G., *An Introduction to Matrices, Sets and Groups for Science Students*, Dover Publications, Inc., New york (1986).

Yourgrau, W., and Mandelstam, S., *Variational Principles in Dynamics and Quantum Theory*, Dover Publications, Inc., New York 1979).

Dynamics

Becker, R. A., *Introduction to Theoretical Mechanics*, McGraw-Hill Book Company, Inc., New York (1954).

Byerly, W. E., *An Introduction to the Use of Generalized Coordinates in Mechanics and Physics*, Dover Publications, Inc., New York (1965).

Kilmister, C. W., *Lagrangian Dynamics: an Introduction for Students*, Plenum Press, New York (1967).

*Ramsey, A. S., *Dynamics Part I*, Cambridge University Press, Cambridge (1951).

*Routh, E. J., *Dynamics of a System of Rigid Bodies*, Dover Publications, Inc., New York (1960).

Whittaker, E. T., *A Treatise on the Analytical Dynamics of Particles and Rigid Bodies*, Cambridge University Press, Cambridge (1961). This is a classic work that goes well beyond the level of the present book. It is, nonetheless, well worth consulting to see what lies ahead!

Relativity and Gravitation

*Einstein, A.., *The Principle of Relativity*, Dover Publications, Inc., New York (1952). A collection of original papers on the Special and General Theories of Relativity.

Dixon, W. G., *Special Relativity*, Cambridge University Press, Cambridge (1978).

French, A. P., *Special Relativity*, W. W. Norton & Company, Inc., New York (1968).

Kenyon, I. R., *General Relativity*, Oxford University Press, Oxford (1990).

Lucas, J. R., and Hodgson, P. E., *Spacetime and Electromagnetism*, Oxford University Press, Oxford (1990).

*Ohanian, H. C., *Gravitation and Spacetime*, W. W. Norton & Company, Inc., New York (1976).

*Rindler, W., *Introduction to Special Relativity*, Oxford University Press, Oxford, 2nd edn (1991).

Rosser, W. G. V., *Introductory Relativity*, Butterworth & Co. Ltd., London (1967).

Non-Linear Dynamics

*Baker, G. L., and Gollub, J. P., *Chaotic Dynamics*, Cambridge University Press, Cambridge (1991).

Press, W. H., Teukolsky, S. A., Vetterling W. T., and Flannery, B. P., *Numerical Recipes in C*, Cambridge University Press, Cambridge 2nd edn (1992).

Waves

Crawford, F. S., *Waves*, (Berkeley Physics Series, vol 3), McGraw-Hill Book Company, Inc., New York (1968).

French, A. P., *Vibrations and Waves*, W. W. Norton & Company, Inc., New York (1971).

General reading

Bronowski, J., *The Ascent of Man*, Little, Brown and Company, Boston (1973).

Calder, N., *Einstein's Universe*, The Viking Press, New York (1979).

Davies, P. C. W., *Space and Time in the Modern Universe*, Cambridge University Press, Cambridge (1977).

Schrier, E. W., and Allman, W. F., eds., *Newton at the Bat*, Charles Scribner's Sons, New York (1984).

25157631R00107

Made in the USA
Columbia, SC
02 September 2018